Pilot Error
A professional study of contributory factors

Pilot Error

A professional study of contributory factors

Edited by Ronald Hurst, AMRAeS

Crosby Lockwood Staples   London

Granada Publishing Limited

First published in Great Britain 1976 by
Crosby Lockwood Staples
Frogmore St Albans Hertfordshire AL2 2NF and
3 Upper James Street London W1R 4BP

ISBN 0 258 97072 3

Foreword

Sir Peter Masefield, MA, FRAeS, FCInstT, Hon FAIAA (USA), Hon CASI (Canada)

An immense amount of effort, analysis and discussion has been devoted, over the years, to every aspect of the manufacture, the operation and the economics of air transport. As a result, commercial air services are now, in relative terms, safe – and getting safer. In this, as in many other aspects, air transport now compares well with all the more mundane forms of travel. Statistics show, indeed, that on average, an air traveller endowed with inordinate longevity (having doubtless drained the Elixir of Eternal Youth) could continue to fly on scheduled air services, day and night, for some 200 years before, statistically speaking, his 'number came up'.

That is, however, only one side of the picture. The fact is that accidents do happen – however remote they may seem to the individual traveller, and however much the odds are against any particular person being involved in one of them.

In recent years, over the field of scheduled air transport, there have been an average of about 31 fatal accidents a year in which, annually, an average of some 950 passengers have been killed out of now some 440 million individual annual air journeys. The odds are, therefore, around half a million to one against any passenger being killed on any one flight.

When we come to examine those accidents that do happen, however, we find that about half of them – and half the number of fatalities – have been (and continue to be) ascribed to 'human error'.

The truth is that in some degree or other every accident is a result of human error, whether that error be direct – as, for instance, a pilot error on landing – or indirect, in as, for instance, an insufficiently thought-out design feature which has led to a catastrophic failure of some component of an aeroplane or of its systems. Every mechanical process is potentially vulnerable to some human mistake or misjudgement, however far back along the line that failure may have been.

The complexities are enormous, yet on their understanding and resolution depends the further substantial improvement of the already high safety standards that exist in air transport and the enhancement of the less good safety record of other branches of aviation.

An important step towards the understanding of the underlying factors which make up air safety, expressed in human terms, is contained in this book. I commend it to all those who are concerned not only with aviation and its development as a servant of mankind – whether for transport, for defence or for pleasure – but to all those who are concerned with the human aspects of technological development in its multifarious forms. Most of all, I commend it to that dedicated band of hard-working technocrats – the modern airline pilots.

Aviation is an especially suitable subject for study in these terms because it is so much of a front-runner in design criteria, in development and manufacturing standards and in what has come to be termed a 'fail safe' philosophy.

Aircraft operations, which involve the combined and fundamental problems of large bodies being propelled at great heights at high speeds and then brought back to earth again at specific times and finite places, are, at the same time, both complex and unforgiving of error. If an accident *can* happen, sooner or later it *will* happen.

At the same time, very few accidents happen from a single cause. Almost invariably they are a combination of causes. Foresight, skill and experience have been used to guard against specific failures. When they do occur they are usually signalled and safeguards come into action. Accident investigation shows that the disaster still occurs when there is a compounding of factors – mechanical and human – the last of which goes beyond the provisions for which previous experience and all reasonable foresight has allowed.

Over the years, accident investigation has brought to light potential hazards. As each one is exposed, that hole in the fabric of safety is stopped up. In the end what is left is the human element because, in the last analysis, however perfect the machine, a man can make a mistake.

There are prospects, however, of developing safety techniques in air transport eventually to make it the safest means of travel extant – and many times safer than the roads.

The trend in aviation, as in most other aspects of technological development, is unswervingly towards automation; towards the elimination of the unpredictable. And, because most aviation accidents occur either on landing or at take-off, 'auto-land', followed by 'auto-take-off' procedures, not just double-banked but treble-banked against failure, are gradually being introduced. Then the human element can be used in what is probably its most efficient role – that of a monitor and control guardian.

The various chapters in this book, each written by a specialist on his subject, combine to illuminate the problems and the prospects that line the road to still greater safety in the air. In a sense they illuminate too the fact that, underlying the history of safety endeavours, there exists a clear undercurrent of natural resentment among active pilots, on whom ultimate responsibility inevitably rests, that the blame for pilot error (human frailty) is always lurking just around the corner and has been so often expressed.

The human element will always be there in any man-made device or operation; but at the back of almost every charge of 'pilot error' there exists one, or both, of two more serious charges:

Too difficult,   or   *Fatigue.*

For safety, critical operations must be made easy and must be made forgiving of mental aberrations. For safety, the problems of fatigue must be eliminated: easy to say – difficult to achieve. They stem back to the elements of the business – right to the drawing board and the flight duty roster.

Consideration of the many facets of these issues can all contribute significantly to the understanding of requirements for still greater safety in the air.

This book is one of the most important contributions to air safety that has been compiled because it sets out in a more comprehensive way than has been gathered together before the factors which in the long run will lead to that desired end – the elimination of the words 'pilot error' from any report.

Peter Masefield
1 July 1976

Dedicated,
with respect, to
Captain James Thain

Calamity comes like the whirlwind
PROVERBS 1.27

Contents

Preface

This book is an endeavour to study one of aviation's most challenging problems through the eyes of some of those most intimately concerned, and to present an integrated picture of facts and attitudes which may be of value to researchers in the future.

The enigma of pilot error must yield in time to the same objectivity of analysis which is directed to the problems of aerodynamics, power plant or aircraft operations. Yet, while this is the common orientation of research, it is clear that the input of essential information in this sphere suffers considerably from its admixture of human bias, reticence and myth; and that – unlike the approach to purely 'technological' problems such as those mentioned above – pilot error investigations have too often been bedevilled by treatment which has been subjective, rather than scientific.

To such effect is this true that the emotive term 'pilot error' is now synonymous in the minds of press and public with an instant verdict of 'guilty' against the pilot(s) concerned – a reflex which is consolidated by the natural reluctance of the air transport industry to publicise its additional and sometimes insoluble difficulties. It is, however, a reluctance which is increasingly pervasive and wholly damaging in effect.

No other technology would tolerate such an obvious hiatus in the pattern of its logic, and, in suffering this situation, aviation (and aviators) have inevitably suffered the consequences of inadequate vital communication, inhibited remedial studies and development and, occasionally, monumental human injustice. This book is dedicated to such a victim.

It is timely, in this light, to make the case for a more realistic and more general understanding of the pilot's task within the current state of development of the air transport industry, and it is believed that this end is best served by the course adopted here: namely, by placing the pilot in the context of his environment. It will be seen that, far from exercising his skill in isolation, he is but an instrument responding to external as well as subjective pressures. It is a fact that, where allegations of pilot error have been made, 'pilots have rarely survived either to offer or to contest the evidence'; in identifying many of these pressures, this study must in some measure speak for them.

My most sincere thanks are due to the hardworked team of contributors to this book. Firstly, to Captain Peter Bressey, who gave so unstintingly of his expertise, who guided me patiently through so many difficulties while this work was in preparation, and who provided a chapter of exhilarating candour; to Dr Martin Allnutt, of Farnborough's Aviation Medicine Division, for his fascinating analysis of the human factor and its relation to pilot error; to John Allen of Hawker Siddeley whose inspiration it has been a privilege to share; and to Philip Martin, for his equally candid account of the world of Air Traffic Control. A special word is due to my contributor-at-long-range, Captain Arne Leibing of Sweden's Civil Aviation Flight Safety Department, for an unusual and important light on a fundamental matter of air safety; and to Captain Norman Price for a chapter which will surely continue to provide thought for every pilot, passenger and operator.

I am glad to place on record that the above thanks are offered, not simply to 'authors', but to friends, and to stress that the views expressed are those of the individuals concerned. In no way do these views purport to represent the orientation or policy of any official body, private organisation, or other person, unless such attribution is made in the text.

Ruthlessly, as editors do, I pressed other friends into my service: John Cutler, AFRAeS, who spent many weary and uncomplaining hours on the illustrations; and Aubrey Wilson, whose practical help and enthusiasm sustained this book from the moment of its conception.

Finally – and because it so clearly illustrates the atmosphere of emotional confusion which surrounds the phrase 'pilot error' – it is important to draw attention to the reservations of those in aviation who felt that this theme was not a proper subject for publication; and who, on those grounds, chose not to participate in the making of this study.

Ronald Hurst
Dean Gate
Gallows Tree Common
Oxfordshire, England
February 1976

1

Pilot's View
P. E. Bressey

Every pilot is a human being and therefore fallible.

> British Airline Pilots Association submission to the Cairns Committee on Accident Investigation, 1961.

The causative factors which precipitated the error are usually overlooked.

> A. L. Steinberg, Douglas Aircraft, 1964.

It is a grim axiom of the aviation community that the pilot is the first person to arrive at the scene of an air accident; and it may be assumed that this knowledge provides pilots with one of the most basic, most urgent, and most compelling incentives for professional vigilance.

Yet pilot error has been recorded as the major cause of the largest number of air accidents to date – a statistical fact which lays the responsibility for more than 55 per cent of these occurrences squarely on the rigorously selected, highly trained, disciplined and vocationally dedicated airmen who flew the unfortunate aircraft.

It should occasion no surprise therefore, that their colleagues have come to despise a judgement which they believe has become increasingly and demonstrably inadequate as an objective scientific finding; that in the light of literally generations of knowledge and personal experience of operating conditions and techniques, airline pilots should instinctively resent and reject the dubious pilot-error verdict; or that they should call into question not only the juridical, but even the ethical basis and motivation of such a verdict against a defendant who has but rarely survived either to offer, or to contest, evidence.

It is no function of this chapter to suggest that the airline pilot possesses less than the human quota of fallibility, or that this characteristic will remain inflexibly disciplined throughout the 20 000–30 000 flying hours of the average commercial pilot's career. It is clear, however, that the pilot-error verdict does much more than merely satisfy the legal requirements of air accident investigations. Additional, and wholly inadmissible connotations reside in a ruling which, by definition, appears to absolve all others concerned with the design, manufacture or operation of the aircraft from any contributory responsibility for the accident; and which, again by definition, apportions no culpability to the cumulative effect of stresses which may represent any intolerable mix of human, mechanical, or environmental factors.

It is the common purpose of the contributors to this book – each one an acknowledged professional representing a key facet of modern aviation – to examine these factors; and in so doing to begin with a pilot's view in order to repair the invariable omission of this witness from far too many air-accident investigations in the past.

The situations and pressures illuminated here stem from the considered impressions of a senior airline captain. Many of these situations and pressures represent the recorded ordeals of other pilots. Many are grounded in the personal experience of the author in the course of three decades of commercial aircraft operation. The sum of this knowledge must strengthen the challenge to future pilot-error verdicts wherever they are manifestly in conflict with true accountability. Such conflict, as will be seen, is a matter of historical fact.

The frame of decision

It has been said of pilots that they are good differentiators, but not good integrators; an assertion which recognises the possibility that too high an escalation of demand on the pilot may cause him to make the wrong decision. It may appear surprising that so skilled a technician should be regarded as being of dubious reliability in this respect, since, in industry and commerce, decision-making is a long-established index of managerial capacity. The parameters of the art are well documented and there is wide acceptance of its conventions.

Typically, the claim for sufficient time in which to consider any major business decision is signalled by the use of cliches. The logic of the situation is apparent and it is unnecessary to labour the point beyond '... I must have time to think about this/I'll ring you back/consult my partners/sleep on it and let you know in the morning ...'. It would therefore be regarded as a grotesque aberration if a decision on the disposition of equipment worth several millions of pounds, and on the safety of two or three (or latterly four) hundred people – including the executive from whom the decision is requested – were to be insisted on in the space of *three to five seconds.*

Yet this is the decision frame for the airline pilot making an instru-

Table 1 Hull losses (excluding sabotage)
☐ Take-off accidents (% of total)
▨ Approach and landing accidents (% of total)
(Courtesy C. Lawson White, Senior Operations Officer, IATA)

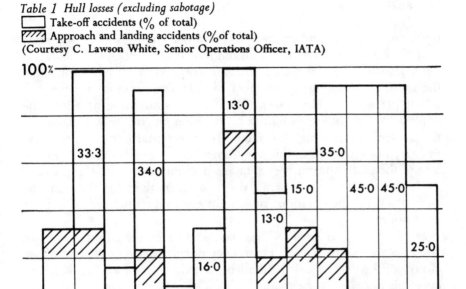

This analysis of more than ten years of operations shows that over half of the accidents occurred during the approach and landing phases.

Flight (23 January 1975 and 24 January 1976) records fatal approach/landing accidents for the years 1972–75, which may be summarised as follows:

1972	21 accidents	public transport and
1973	23 accidents	executive jet aircraft
1974	10 accidents	
1975	10 accidents	scheduled passenger flights

The 1975 analysis states that 'one in two fatal airliner accidents occurs during approach and landing' and refers to 'more than eighty' such accidents in the previous five years. The 1976 analysis, again, points out that the incidence of approach/landing accidents remains at 'more than half' the total number of accidents listed.

ment approach to land in marginal weather conditions. In the most graphic and basic terms, he is sliding down an electronic banister in the sky with some 200 tons of aircraft strapped to his posterior while the ground is coming up to meet him at an approximate speed of 230 ft per second. He knows that if he sees anything at all of the approach lights before reaching his decision height, he will have less than 5 seconds in which to make his operational decision; namely, to carry on with the landing, or to overshoot; and he knows – in the face of the stark alternative – that his decision *must* be the right one.

He cannot slow the aircraft down to gain time for thought; he cannot put off the decision; he cannot even take time out to consult his co-pilot.

In this case he has less than 5 seconds. It will be seen that there are other occasions – such as an emergency during the most critical phase of take-off – when he will be denied even that timespace in which to make his decision, *and act on it.*

The time element is particularly crucial during the take-off, and the approach and landing phases of flight; a fact, illustrated in Table 1, which is reflected in the remarkably high proportion of accidents occurring during these two phases. Thus, their vital nature is acknowledged. It has already been remarked that the human mind does not take kindly to making important decisions in a hurry. What kind of situation then, can impose the urgency and the associated psychological stress at those times?

The aborted take-off; theory and practice

Probably the most difficult 'decision in seconds' which ever faces the air-line pilot is an emergency during the critical phase of take-off. In order to understand what is meant by this term it is necessary to delve into the mysteries of the *Flight Manual* – that work which, in effect, is the pilot's Bible. Here, in page after page of graphs and tables, guidance is offered to cover every phase of flight for every conceivable combina-tion of operational factors – aircraft weight, air temperature, surface wind, runway length, number of engines inoperative ... and so on.

It is unnecessary to detail the mathematics of the performance cal-culations governing take-off in this chapter. Suffice it to say that every multi-engined commercial aircraft must be able to take off after an engine has failed provided it has reached a certain speed at the moment the failure occurs. This speed, of course, varies with the air-craft weight, and also with the runway length and altitude; but it is selected for each take-off so that if an engine failure takes place *after* the 'decision speed' (V_1) has been reached, the momentum is such that the power of the remaining engine(s) will accelerate the aircraft to its minimum take-off speed and then into the air, so that a safe take-off may be accomplished. The decision speed is also chosen so that if engine failure takes place *before* V_1, the aircraft can (theoretically, and on a dry runway), be brought safely to a halt before reaching the far end of the runway.

It is this phase of the manœuvre (known as 'accelerate–stop'), which presents the greatest difficulty in practice. 'To go or not to go' may seem to be a simple enough decision; but if the emergency arises *just before* the V_1 speed is reached, the human time lag in assessing the circumstances and *the nature of the action which must follow*, assumes enough importance to become the determining factor in the manœuvre's success.

The *Flight Manual* figures, however, are obtained by the manufacturer's test pilots, working under optimum conditions, i.e. from an ideal runway and forewarned with the knowledge that they are going to have to stop the aircraft from a speed just below V_1. Since human time lag tends to be reduced with practice, a test pilot who has accomplished the accelerate–stop drill several times can easily demonstrate more rapid response than the airline pilot, who may only have done one drill in a simulator in the past six months, and who may never have carried out the exercise at high speed in the aircraft itself. The contrast between the simulated and the real emergency is further heightened (a) by the fact that because this manœuvre is extremely expensive in both brake and tyre wear, rejected take-offs are never given in training at speeds above 50 per cent of the V_1 speed; and (b) – even more important – by the fact that the airline captain is not *expecting* to have to abort the take-off.

The optimum figure for human reaction time for the simplest task, in which no decision is required, is 0·2 seconds, and for even simple tasks it takes about 20 per cent longer to respond with the feet than with the hands. Reaction time, of course, is not only a function of the body member used, but also of the complexity of the response required, and of the nature of the stimulus. For very complex stimulus and response conditions such as are now under consideration, reaction times can be as long as 3 or 4 seconds.

The test pilot conducts his (forewarned) accelerate–stop in the most efficient manner possible. His heels are raised off the floor, his toes are poised on the brake-pedals, and his first action is to apply these – expense no object, as line pilots dryly observe.

The airline captain, however, when faced with the sudden decision to abandon the take-off, reverts to his fixed habit of stopping the aircraft as he normally does; which means that the most important aspect – applying the brakes – is his *last* action after closing the throttles, selecting spoilers or lift dumpers, and reversing thrust. Furthermore, he has been trained to make all take-offs with his toes well clear of the brake-pedals to prevent inadvertent application when operating from rough, bumpy runways.

If the accelerate–stop is being made from this sort of runway (and there are plenty of them about) the vibration transmitted to the instrument panel by the pounding of the nose-wheel may well make it extremely difficult for the co-pilot (responsible for calling V_1) to read the air-speed indicator accurately. This, too, can cause problems.

Engine failure; power? or no power?

It will have been noted that although the original reference was to 'an emergency during the critical phase of take-off', the certification tests are made simulating an engine failure. The calculations are thus made on the assumption that, in the interval between the engine failing and the first application of brakes, the aircraft continues to accelerate at the engine-out rate. Yet if the 'engine failure' recognition signal received by the pilot is a firewarning light plus bell, he has no time to wonder whether the warning is real or spurious – he will attempt to stop. Even if the fire-warning is real, the engine may well continue to give full take-off thrust for the next few seconds, thus maintaining the all-engines acceleration rate during the recognition, decision and reaction time periods, known also as the 'transition distance'. In fact, records show that less than half of all the abandoned take-offs on the line result from engine failures. Plainly, there are many other malfunctions which the pilot is unwilling to take into the air with him.

All these elements add up to one unpleasant fact: that the airline pilot, faced with a sudden crisis just before V_1, is very unlikely to stop his aircraft within the flight-test distance. And as aircraft become larger and heavier and faster, more and more runways round the world become 'critical runways' for take-off. In consequence the take-off weight must be reduced so that the 'Flight Manual distance' needed for an accelerate–stop is just equal to the runway distance actually available.

At the V_1 speed for a modern jet, the aircraft is travelling down the runway at anything from 220 to 280 ft every second, and the extra margin of time needed by the airline pilot to recognise, decide, and react to an unexpected emergency is very costly in terms of 'transition distance', and so of total runway distance used. In a simulator study carried out in the USA more than 40 experienced captains were presented with a sudden, unexpected emergency at 7 knots before V_1. The test showed that, on average, an unprepared line pilot, using normal airline techniques, needed an extra 70 per cent of 'transition distance' compared with the forewarned test pilot using certification techniques.

Solutions: cost: safe landing speed

Like so many other safety aspects of commercial aviation, the solution to this problem is expensive. Either a large number of runways all over the world must be lengthened, or the airlines must agree to reduce take-off weights by a considerable amount when operating from these runways. Pilots have been pressing both nationally and internationally to obtain modification of the existing take-off performance specification, and many of the facts quoted above have been taken from a 20-page *Working Paper* submitted by the International Federation of Airline Pilots Associations, IFALPA, to the Airworthiness Committee of the International Civil Aviation Organisation, ICAO, as long ago as November 1970. As with so many other 'grey areas' of aviation's problems, however, the fight for safer practice continues, still unresolved.

Table 1 also shows that the number of accidents that occur during the approach and landing phase are almost twice those that occur during take-off. The shortage of time has already been mentioned as a vital factor of the approach in marginal weather conditions, but even in good weather the large number of variables, plus the almost endless permutations of these variables, combine to make this the most difficult, and therefore the most error-prone phase of flight.

It is relevant in this context to observe that man is a two-dimensional animal, whose mind and body had, until the beginning of this century, evolved to cope with problems in two dimensions; and that he has neither the physical equipment nor the inherent knowledge for making three-dimensional judgements, especially at high speed. In the period between the two World Wars, the internationally famous aviation writer, C. G. Grey, was continually preaching the necessity for aeroplanes '... which would land slowly and not burn up' since even in those early days, the landing accident was the most frequent. His pleas met with little success inevitably, since slow landing speeds must be paid for either in weight, i.e. in reduced payload, or in reduced high-speed performance. In the event, neither manufacturers nor operators – each preoccupied with their respective commercial pressures – will lightly accept such penalties.

That well-known author of aviation novels, the late Nevil Shute, was even more outspoken on the subject. Writing about his early days as an aeroplane designer in the mid-1930s, he declared that 'Experience has taught me one sad fact – that you can't sell safety. Everybody pays lip service to the safety of aeroplanes, but no one is prepared to pay anything for it.'

A scathing generalisation; but it is a fact that with the outstanding exception of the VC10, every civil air transport designed in the past 30 years, both in this country and in America, has had a higher landing speed than the aircraft it was built to replace; and the VC10, despite its popularity with both pilots and passengers, failed to sell in large numbers because of its high operating costs.

Pilots of modern airliners, then, are committed, among other things, to fast landing approaches. The reader is now invited into the cockpit in order that he may share some aspects of this experience.

The runway: seeing and believing

The path in space which a pilot desires to follow during an approach may be regarded as the intersection of two reference planes – the vertical plane through the centre of the runway, and the 'glide path'. In order that the approach may be stable and accurate, the pilot needs to know *continually* the values of six variables:

1 and 2. the displacement relative to each reference plane,
3 and 4. the track heading (or rate of closure), relative to each reference plane,
5 and 6. the rate of change of the track heading (rate of turn) relative to each reference plane.

In correcting an error in either plane, the pilot's task is continuously to match heading (the direction in which the nose of the aircraft is pointing) against displacement, so that at the moment at which the displacement eventually becomes zero, the relative track heading and the rate of turn have also become zero. Any suspicion on the part of the reader that all this sounds extremely complicated is entirely justified; particularly since corrections must be made in both planes at the same time.

Figures 1 and 2 show that the visual clues available for judging the

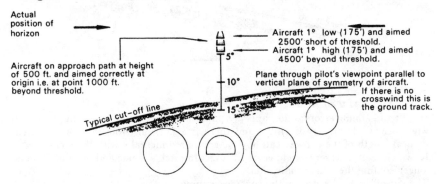

Fig. 1 Pilot's view of runway: vertical extension

Height and aiming errors as they appear at approx. 2 miles out on the approach, when only the runway is visible and there is no horizon. Note that the only difference in shape of the three images is that the convergence of the edges becomes greater as the height is reduced. Apparent length of runway cannot be used as a height indication because the length seen depends on the visibility.

The cut-off line is formed by the cockpit coaming.

Distance from approach path origin is 10 000 ft in all three cases.

Approach path angle is 1 in 20.

Visual range is 11 000 ft, i.e. 2000 ft of runway is visible.

Airspeed and angle of approach assumed to be constant, i.e. rate of descent is constant. No crosswind, bank or heading error.

(Courtesy Dr E. S. Calvert, late of the Royal Aircraft Establishment, Farnborough)

approach relative to the two planes vary enormously. It is comparatively easy for the experienced pilot to decide if he is on the extended centre-line of the runway, or off to one side. But displacement above or below the inclined plane containing the optimum glide path merely results in vertical extension or compression of the perspective picture of the runway (Fig. 2), and the only way the pilot can tell if he is on glide path is by comparing the actual view of the runway he sees with an 'ideal image' he carries in his mind – an image which has been impressed there by training and practice.

Because the differences between the ideal image, representing zero vertical displacement, and an image representing quite a large vertical displacement, are small, keeping on glide path is by no means easy.

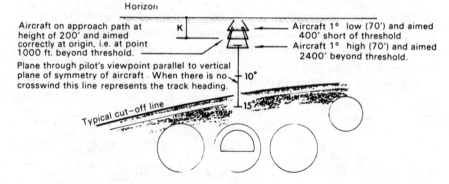

Horizon

Aircraft on approach path at height of 200′ and aimed correctly at origin, i.e. at point 1000 ft. beyond threshold.

K

Aircraft 1° low (70′) and aimed 400′ short of threshold

Aircraft 1° high (70′) and aimed 2400′ beyond threshold.

Plane through pilot's viewpoint parallel to vertical plane of symmetry of aircraft. When there is no crosswind this line represents the track heading.

10°

Typical cut—off line

15°

Fig. 2 Pilot's view of runway: vertical compression

Height and aiming errors as they appear at approximately ¾ mile out on the approach when nearly all runway is visible and there is a horizon. Note that the apparent vertical length of the runway can now be compared mentally with the ideal length, K, which represents zero height error. Changes in pitch attitude and head position no longer confuse the height indication.

The cut-off line is formed by the cockpit coaming.

Distance from origin is 4000 ft in all three cases.

Approach path angle is 1 in 20.

Visual range is as in Fig. 1, i.e. 11 000 ft; 8000 ft of runway is visible.

Airspeed and angle of approach assumed to be constant, i.e. rate of descent is constant.

(Courtesy Dr E. S. Calvert, late of the Royal Aircraft Establishment, Farnborough)

The unhelpful jet

A large part of a pilot's initial training is devoted to teaching him how to make these three-dimensional judgements, and the smooth and accurate corrections which are necessary to avoid continually over-shooting the desired path (in either plane). But his already difficult task in maintaining an accurate glide path has been made even harder by the introduction of the jet engine.

The propeller-driven aircraft had an almost instantaneous reaction to an increase of power – the increased airflow over the wings from the speeded-up propellers produced an immediate increase in lift. How-ever, not only is the jet engine slower to accelerate from small throttle openings than the piston engine, but it has no direct effect on the airflow over the wings. Not until the whole mass of the aircraft has been accelerated due to the increased thrust will any increased wing-lift result. Thus any deviation below the glide path should be instantly recognised, and corrective action initiated immediately; yet this is the very area in which visual clues are at their weakest.

By the time the pilot recognises that he is below the desired approach path, a large application of power is necessary to regain it, which may result in the aircraft going above the glide path, and possibly setting up a porpoising effect.

Ever since the first introduction of the jet into airline service, pilots have been pressing for a precision approach aid providing guidance in the vertical plane (such as Instrument Landing System (ILS)) to be installed on all runways used by jet aircraft. *After two decades of constant lobbying, less than one in three of such runways is so equipped,* and undershoot accidents continue to occur with distressing frequency – often in quite good weather, when the cause is claimed to be 'obviously pilot error'.

Wind and weather

Except in rarely encountered calm conditions, the effect of wind must be taken into account, since unless the wind is blowing directly down the runway, the nose of the aircraft will have to be pointed off to one side (into wind) to counteract the effect of drift. The strength of the wind normally decreases as the aircraft approaches the ground, so the pilot will have to make continuous adjustment for this; but under certain conditions 'wind shear' (a very rapid change of wind strength and direction) may be suddenly encountered at low altitude.

This phenomenon is difficult to forecast, but its effect on the approach path of an aircraft can be highly dangerous. Turbulence due to strong and gusty winds or to high temperatures on the ground add to the pilot's difficulties, as do most kinds of precipitation.

The falling rain-drops from a light shower, which hardly diminish the distance an observer on the ground can see (the 'met. visibility'), have a much greater obscuring effect when battered against an aircraft windscreen by a 160 m.p.h. slipstream. Any motorist who has walked to his car during a snow shower and then commenced to drive away has experienced the immediate reduction in forward visibility due to speed, whenever precipitation is present. The motoring analogy can also be applied to the problem of controlling a moving vehicle in fog; the prudent motorist reduces his speed so that he has time in which to recognise the visual clues as they appear out of the murk and then take the appropriate action. If he runs into a thicker patch, he slows down to an even more cautious speed.

For the airline pilot on his approach to land in fog, this problem is complicated in two ways: he has an extra dimension (the vertical) to cope with, and he cannot slow down below his minimum approach speed of 220 to 250 ft per second. In order, therefore, that he should have reasonably standardised clues on which to base his visual judgements, landings both by day and by night are made only on Category II runways when the visibility is less than half a mile. Such runways are not only equipped with ILS so that the aircraft can be accurately positioned for the transition from instrument to visual flight, but also with the standard 3000 ft of centre-line and cross-bar approach lighting plus full runway lighting. Both approach and runway lights are always turned on for low visibility landings by day as well as by night.

For the vast majority of present-day civil airliners, the approach will be made on the ILS down to decision height – normally between 100 and 200 ft above runway level. By the time this height has been reached, the pilot must have made his decision – to overshoot, or to continue the approach and landing 'solely by visual reference' to the lights. To prevent pilots attempting to land in dangerous conditions, 'operating minima' in terms of decision height and Runway Visual Range (RVR) are listed in the *Company Operations Manual* for every Category II runway, and a pilot may not even commence an approach to land until the RVR is promulgated as above his company minimum for his aircraft.

How much runway? The visual segment

RVR is an assessment of the distance along the runway which a pilot should be able to see from an eye height of 16 ft, and is normally measured by a transmissometer, a photo-electric device which measures the obscuration of the atmosphere just alongside the runway. The transmissometer reading is then converted into RVR, taking into account such factors as the intensity setting of the lights, and whether it is day or night. The end product of all these calculations is then passed to the pilot.

Ideally, he should also be provided with Slant Visual Range (SVR), which is the distance along the approach path that the lights can be seen. At the time of writing, however, no equipment has yet been developed capable of providing this information; so the pilot, having been told that the RVR is above his minimum, must carry on down to the decision height, hoping that by the time he arrives at this point on the approach he will (in the pilot's jargon) 'be carrying his minimum visual segment'.

The visual segment is the length of the approach light pattern which is in the pilot's view at any one instant of time (Fig. 3). The furthest

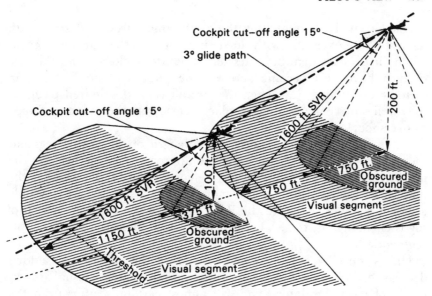

Cockpit cut–off angle 15°

3° glide path

Cockpit cut–off angle 15°

Fig. 3 Visual segment
The aircraft is approaching down a 3° glide slope and has a cockpit cut-off angle of
15°. At a height of 200 ft the first 750 ft of the ground is obscured, so that the
first approach lights visible are 750 ft ahead. At a ground speed of 250 ft/sec the
minimum length of visual segment required is 750 ft. To obtain this visual segment
a Slant Visual Range of 1600 ft is required.

light visible is determined by the fog (SVR), and the nearest light by
the angle of downward view from the cockpit during the approach –
the 'cockpit cut-off'. This is of the order of 15° below the horizontal
for most modern airliners. *The minimum visual segment is then the shortest
length of the approach lights which the pilot must have continuously in view
in order to be able to control his aircraft safely.* This particular definition
has been the cause of continual argument and reappraisal amongst
pilots since it was first introduced over 20 years ago. The most generally
accepted figure was '3 seconds-worth of ground-speed', but this is now
suspect for a modern jet.

The moving platform

Orientation is much more difficult in the air than on the ground,
since the observations are made from a framework which can move
bodily about all three axes, and at the same time, rotate about all
three axes. Thus, the feed-back of rate information to a pilot is a complicated process. In scientific terms, every time a pilot makes a visual

correction, he is '... matching the displacement from a desired path against a rate of closure with the desired path, so that, at the moment when the displacement becomes zero, the rate of closure and the rate of turn have also become zero. If the pilot does not succeed in 'zeroing everything' just when the aircraft is on the desired path, he may well replace the original displacement by another – and be forced to overshoot!

When the fog is homogeneous, the length of the visible segment increases as the aircraft continues the approach. This is because the amount of obscured ground below the cockpit cut-off decreases as the descent continues until, at the moment of touch-down, the visible segment is almost equal to the RVR. However, if a thicker patch of fog is encountered, the SVR decreases, and so does the pilot's visible segment.

Unless the pilot has been well briefed on the effects of a sudden decrease of SVR (which is impossible to reproduce on the visual attachments of most simulators), he will almost certainly suffer from the illusion that he has inadvertently raised the nose of the aircraft – thus reducing his cockpit cut-off angle and obscuring the nearest two or three lights of the approach light system. The natural reaction is then to press forward on the control column, to regain the 'original' attitude. This of course increases the rate of descent, and thus increases the chances of undershooting the runway. Many undershoot accidents and less serious incidents have been caused by pilots misinterpreting ambiguous visual clues in this manner.

As with so many other contributory causes to 'pilot-error' accidents, a solution to the problem – in this case of undershoot accidents *in all weathers* – is readily available, but expensive. The Head Up Display (inevitably shortened to HUD) has been fitted to the advanced military aircraft of many nations for the past decade, and over the last eight years, airline pilots have intensified their efforts to persuade the airlines to equip each new addition to their fleets with a form of HUD. The original decision to install HUD in the first production model of the Concorde for British Airways has not in fact been put into practice.

The HUD system

Basically, the HUD consists of a sheet of heavy glass, chemically treated so that electronic symbols can be displayed on it. The glass hangs down inside and in front of the pilot's windscreen, and since it is translucent, the pilot can look through it and observe the outside 'real world'. Almost any combination of instrument readings, such as airspeed and altitude, plus information about the 'outside world' such as the horizon, the aiming point, and even the outline of the runway itself, can be included in the display.

The input is programmed through a computer, a cathode ray tube, and a series of lenses, and is finally projected symbolically, without parallax, onto the glass screen. Since the HUD is gyroscopically controlled and focused at infinity, the 'instrument horizon' and the real horizon will coincide when the latter becomes visible, and so will the aiming point and even the runway outline. The pilot can thus make a gradual and relaxed transition from instrument to visual flight without having to refocus his eyes, and in the event of any momentary diminution of the visual clues he will continue to receive uninterrupted accurate information from the HUD.

If an overshoot is required, the necessary flight director information can readily be programmed into the computer, to be instantly available – for example, whenever the throttles are advanced to the maximum thrust position. All this, however, is pie in the sky as far as current operations are concerned, and on far too many runways today the pilot is still left to judge his approach with no aids beyond his own skill and experience.

The Fatigue Problem

Most humans like to tackle a particularly difficult task when feeling fresh and rested, and pilots are no exception. However, the most severe calls on a pilot's skill and judgement come at the end of the flight, and the last landing of a number of sectors may well have to be made when he is feeling tired, if not fatigued. It is a vexed word, 'fatigue'; and in all its aspects it has probably caused more argument and ill-feeling between airline pilots and non-pilots than any other subject in aviation history.

There is still no definition of fatigue (as opposed to tiredness) acceptable to all interested parties, nor has medical science yet been able to define, in specific terms, what produces fatigue. But in the pilot's view this loose envelope has encouraged accident investigators over many years to use the most liberal interpretation of 'innocent until proven guilty' in the attribution of fatigue as a possible contributory cause.

In the early 1950s there was a marked lack of accommodation suitable for crew rest at many of the refuelling stops on the world's air routes. This led some operators to schedule extremely long duty periods, and in March 1954, Captain Trevor Hoyle, in command of a BOAC Constellation, was making a visual over-water approach to the runway at Kallang Airport, Singapore, having commenced duty in Sydney, Australia *just over 21½ hours previously*. He misjudged his approach slightly, the undercarriage hit the sea-wall marking the runway threshold, and the Constellation broke up and caught fire, with heavy loss of life. The accident report stated, 'The Captain's error cannot be attributed to fatigue of which he was aware.'

Pilots' working conditions have improved since that date, though as recently as August 1972, the British Airline Pilots' Association (BALPA) set up a special Committee to investigate '... the complicated and contentious subject of fatigue', because of 'Members' disquiet with this growing problem'. Four months after the Association established its special Committee, the Civil Aviation Authority of the United Kingdom (CAA) in turn set up a Committee on Flight Time Limitations under the chairmanship of Group Captain Douglas Bader, CBE, DSO, DFC. The BALPA report, which ran to some 45 000 words, was first submitted to the Bader committee before it was published in February 1973. It was given only limited circulation.

The Bader committee's report (which was considerably shorter) was first published in June 1973, and although its recommendations went a good part of the way to meeting BALPA's requirements, it is interesting to note that the fundamental differences on 'fatigue' between the men who fly the aeroplanes and those who do not, still remain.

The first chapter of the BALPA report analyses ten accidents which occurred between 1964 and 1970 in which pilot error was a possible contributory cause. Five of these ten accidents occurred at the end of a long duty night (of which the shortest was 10 hours 35 minutes and the longest 13 hours 53 minutes). The pilots go on to state: 'In such circumstances fatigue must be suspected to have caused lowered awareness and efficiency on the part of the pilot ... Common sense and our own operating experience lead us to conclude that pilot fatigue could have been a possible contributory cause.'

In an obvious reference to this, the Bader report states: 'Our attention was drawn to accidents in which it was alleged that there was a possibility that fatigue may have been a contributory factor ... *We did not accept the argument that fatigue should be assumed to be a factor, unless proved otherwise, in every case where the duty period was a long one or the flight was at night*' [Author's emphasis].

Pilots, again, find the greatest difficulty in equating this statement with the principle of innocent until proved guilty.

The BALPA team define pilot fatigue as '... that degree of tiredness which leads to an impaired ability to fly accurately and to make correct decisions'.

Deliberately low-key, the Bader report states: 'We have come to consider fatigue as a markedly reduced ability to carry out a task ... We found that a number of witnesses tended to confuse tiredness with fatigue, and generally held the opinion that the amount of work should be reduced to a level at which tiredness was the exception. We do not accept this as a valid conclusion'. This difficulty in drawing a line between 'tiredness' and 'fatigue' is typical of the subjective inaccuracies, confusion and misunderstandings which continue to obscure the whole problem of pilot fatigue.

Flying hours and working hours

The popular press made great play with an official statement published by one airline management to the effect that '... on average, our Captains fly less than 500 hours per year.' In terms of industrial employment this figure is so impressive as to demand further examination. Surely, given such conditions, no pilot could justifiably complain of fatigue?

The statistic, however, is of a piece with the drunkard's lamp-post, offering support rather than illumination; for the 'average' figure quoted includes a surprisingly high proportion of training captains (who spend half their working life in the simulator) and management pilots (who spend three-quarters of their time in an office). A more realistic figure for an average line captain, therefore, would probably be nearer 600 hours a year; but what actual duties remain unpublicised?

The airline concerned was a short-haul carrier and, since the introduction of the jet, averaged a flight time for each sector of under $1\frac{1}{2}$ hours. Pilots flying four such sectors in a day will probably only accrue 5 hours of flying time. Flight crews, however, must report one hour before the first take-off; they are on duty during turn-round or refuelling stops, and must complete paperwork at the end of the day which may take another 30 minutes.

Few 4-sector days (or nights) are completed in less than 10 hours of continuous duty, during which time the main meal will probably have been eaten at the controls during flight, headset in place, and so constantly interrupted by communications to and from the ground.

A working pattern such as this shows a ratio of duty hours to flying hours of more than 2 to 1. When stand-by, weather delays, periodic refresher training and so on are taken into account, the ratio between duty hours and flying hours for Captains in the largest UK short-haul airline varies from 2·3 to 3·5 to 1 for Captains, and from 2·5 to 3·7 to 1 for co-pilots, depending on the type of aircraft flown. This means that the 'average' line captain was probably on duty for between 1300 and 1400 hours per year. This is still not a great deal, averaged out over the 47 working weeks; most pilots get four weeks leave a year, plus a further week in lieu of public holidays ... and the office worker may well object that a working week averaging out at some 30 hours appears to be somewhat less than arduous. The pilots retort that it is not the number of hours worked, but the working conditions, and the total lack of regularity in the working pattern, which cause fatigue:

> We think it appropriate to include a reminder that a pilot's duty period cannot readily be compared to a working day of normal employment. He is confined with two, three or four colleagues in a space smaller than the average bathroom. The pressure is lowered to the equivalent of an altitude of 8000 feet or thereabouts. The atmosphere is uncomfortable because the temperature and humidity control is inadequate; cockpit temperatures commonly reach 95°F in the tropics; humidity varies from 5 per cent to 85 per cent.

Problems of the sleep cycle

Discussing the constant irregularities of the pilot's life, the BALPA report points out that the life of an average worker is planned to a regular 24-hour rhythm: in round figures, he sleeps for 8 hours and is then awake for 16 hours, after which he is ready for sleep once more. He normally starts work within 3 hours, and finishes within 12 hours of waking up.

The theory is then developed that each hour of sleep fits a man for 2 hours of wakefulness – if he stays awake for more than 16 hours, his efficiency falls. If he can succeed in sleeping again before the 16 hours have passed, he will 'recharge his batteries' – a 3-hour sleep will fit him for another 6 hours of wakefulness, and so on. However, the report is adamant that (contrary to the opinion apparently held by many airline managements) pilots cannot train themselves to sleep at will, or for as long as should be necessary to fit them for an unusually long duty period. Pilots often go on duty late in the evening, or at the end of a 24-hour rest period – just as they are going into a 'sleep deficit'.

'Pilots are selected on the basis of stable personality profiles, manual dexterity, physical health and, finally, the ability to fly aeroplanes,' states BALPA. 'At no time does any selection process attempt to assess performance under conditions of sleep deprivation, shift change or upset bodily rhythms.'

The report goes on to explain that the short-haul pilot's greatest problem is the day–night–day work pattern (with variations!), 'which is irreconcilable with the body's natural rhythm'. As an example, the report quotes in detail the experience of a Trident pilot, working an apparently innocuous duty schedule of 6 hours on a Sunday night, followed by a 24-hour rest period, and then $10\frac{1}{2}$ hours day duty on the Tuesday. He flew to Belfast and back to London on the Sunday night, getting to bed at 04.15 on the Monday morning (his rest day), and sleeping for nearly 7 hours until 11.00. Since he was due to fly to Belfast again at 06.55 on the Tuesday, followed by Frankfurt and return, he went to bed at 22.00 on Monday. The pilot is quoted thus: 'Unfortunately, I was unable to get to sleep as presumably my body thought it should be up again all night – the result being that I saw every hour go by, 04.00 being my last recollection before being woken by my alarm at 05.30.'

He now had to face a long day having had very little sleep, and survived the first two sectors without incident. But '. . . due to some air-field congestion at the time of our departure for Frankfurt, we were held near the end of the runway for approximately 10 minutes, and I fell asleep. This came as a great shock to me, and I put the blame directly on the integration of these two trips. There may well have been the required time off between them, but in no way do they take into consideration the human clock.'

The Bader committee took note of '. . . opinion which indicated that fatigue is more likely to result from badly planned sequences of work and rest rather than from the actual duration of duty . . . There are some aspects of crew scheduling which are less desirable than others. These include the scheduling of an alternate day/night pattern of work . . . Consecutive duty cycles should be constructed so as to avoid to the maximum extent these and other less desirable practices. We have made a great effort to construct a realistic scheduling framework, and operators will be expected to arrange their schedules within the spirit of our recommendations as well as following the letter.'

Most airline pilots who have read this paragraph remain to be convinced.*

Time zones and the long-haul pilot

So far, only the problems peculiar to short-haul operations have been discussed. For the long-haul pilot, all these are still experienced, but they are compounded by the effect of yet another problem – that of continued time zone change. Combined with the cumulative effect of repeated 24-hour rest periods, the final result is that the long-haul pilot finds increasing difficulty in obtaining adequate pre-flight rest as he travels farther and farther away from base.

The BALPA report states: 'It is our experience that the main effect of simple time zone change is that it impedes the ability to achieve sleep during local hours of darkness. Compounding this disruption of the pilot's sleep pattern are the constraints placed upon his sleep opportunities by his duty pattern ... In the absence of medical evidence to the contrary, we consider that large time zone changes should be avoided or, failing that, a reduction in the permitted duty should be made to offset the performance degradation which disturbed body-rhythms will cause in these circumstances.' (The reduction recommended is one hour whenever the preceding rest station is four or more time zones removed from base.)

The Bader report admits the existence of the effect of time zone changes, but merely recommends: 'Crews should make a conscious effort to plan their activities in accordance with the requirements of their forthcoming duty period irrespective of local time.' No recommendation is made for any reduction in the permitted duty to compensate for the effect of time-zone changes.

Changing the law of the land is always a painfully slow process; and eight months after the publication of their report, the Bader committee had still not completed the work of amending the report where necessary, in the light of comments received. The final task (which will probably take even longer) will be to collaborate with the various legal departments of the government in order to prepare draft

---

* Fortunately, events such as the publication of CAP 371 (see p. 33) may alter this picture, at least for the pilots of transport aircraft registered in the UK. In the absence of international standards for flight and duty times, however (see pp. 187–91), the problems remain for pilots (and passengers) not protected by adequate legislation in their own countries.

amendments to the Air Navigation Act, which will then have to be passed by the British Parliament.

In fact, the official document setting out the effect of the amendments to the *Air Navigation Order* (Civil Air Publication 371) was not published until April 1975 (nearly two years after the Bader report). It was, of course, a compromise between the views of the pilots and those of the operators. Nevertheless, it represented a considerable improvement in pilots' working conditions as compared with those set out in the preceding document (CAP 295).

The scheduled duty period for all pilots operating from their 'home base' is now related both to the numbers of sectors to be flown *and* to the (local) time of starting duty – the more sectors flown, and the later in the day work is commenced, the shorter the duty period. More important for long-haul pilots operating 'down the line', the effects of time-zone changes *and* of the preceding rest period are also taken into account, in considerable detail.

Finally, a Flight Time Limitations Board has been set up, to which disputed schedules may be referred – though at the time of writing, its impartiality has yet to be tested.

The sleeping robot

No one can remember who first coined the saying that too many pilots let themselves be carried away by an emergency to such an extent that solving the emergency takes priority over flying the aircraft. It was true 50 years ago and it is still true now, in spite of vastly improved methods of training and continual re-checking on emergency procedures in the simulator.

The first-ever fatal accident to one of the new generation of 'wide-bodied jets' – a Lockheed 1011 of Eastern Airlines – occurred at Miami, Florida, on the night of 29 December 1972. It is a classic example of crew preoccupation with a potential emergency leading to neglect of what should have been their primary task – maintaining a safe altitude above the ground.

The report of the National Transportation Safety Board (published less than 8 months after the accident), stated that the flight was uneventful until the approach to Miami was commenced. When the landing gear was lowered preparatory to landing, the green light on the instrument panel which would indicate that the nose-wheel strut was safely extended and locked down, failed to illuminate. The Captain wisely decided to overshoot, and was cleared by Miami Control to an altitude of 2000 ft to sort out his problem. The First Officer flew the aircraft manually up to this height, when the autopilot was engaged, together with the altitude hold switch, which would ensure that this altitude was maintained.

At some time during the next 7 minutes, while the crew were attempting to determine whether the nose-wheel strut was in fact locked down (in which case the failure of the light to illuminate would have been due to a minor electrical fault), one of the pilots must have inadvertently leaned against his control column. A pressure of a mere 15–20 lb is sufficient to disengage the altitude hold switch; but since the aircraft was trimmed for level flight, and weather conditions smooth, the resulting disengagement went unnoticed, and the aircraft commenced a gradual 'drift down' from 2000 ft over the uninhabited (and so unlit) Everglades swamps.

After 7 minutes, the crew asked Miami Control for permission to return to the Airport. This was given, and the autopilot heading selector was re-set to turn the aircraft onto the return course, while the crew returned to their task of trouble-shooting. However, once any aircraft is put into a turn, the nose drops and height is lost unless the fore-and-aft trim is adjusted, either by the human pilot, or by the autopilot if, and only if, the altitude hold is engaged. Since it was not, the rate of descent thus accelerated until the aircraft struck the ground, still in the turn.

The Safety Board determined that the probable cause of the accident was: '... the failure of the flight crew to monitor the flight instruments during the final 4 minutes of flight, and to detect an unexpected descent soon enough to prevent impact with the ground. Preoccupation with a malfunction of the nose landing gear position indicating system distracted the crew's attention from the instruments, and allowed the descent to go unnoticed.'

In conclusion, the Board stated: 'It is obvious that this accident ... was not the final consequence of a single error, but was the cumulative result of several minor deviations from normal operating procedures which triggered a sequence of events with disastrous results.'

The altimeter: the mis-reading hazard

Probably the most important operational requirement for any aircraft instrument is that it should be capable of being read instantly and unambiguously under conditions of stress (when errors are most likely to occur). Since pressurised aircraft were first introduced into service, one of the most frequent causes of accidents has been misreading or mis-setting of the altimeter. Unpressurised civil aircraft seldom flew above 10 000 ft, and rates of ascent and descent were limited to 300 ft/minute, for reasons of passenger comfort. The pilot thus had ample time in which to appreciate changes of height, and to keep up with the aeroplane mentally.

Pressurising the cabin doubled cruising altitudes overnight, while rates of climb and descent of over 1000 ft/minute became possible. The introduction of the jet engine raised cruising levels to 35 000 ft, and made possible rates of descent up to 5000 ft/minute. However, the design of the pressure altimeter fitted to the earlier generations of civil jet transports had not caught up with the changing operational needs of the vastly increased range of heights to be indicated, nor with the necessity for immediate appreciation of the exact height indicated during extremely rapid rates of descent.

The original three-pointer altimeter consisted of a dial calibrated from 0 to 9, over which the larger pointer (comparable to the minute hand of a clock) recorded hundreds of feet to the nearest 10 ft, the smaller pointer (the 'hour hand') was geared to record thousands of feet, while a third, stub pointer, recorded tens of thousands of feet. All this information was shown on a dial some three inches in diameter (about the size of the average travelling clock). Particularly when viewed at night, under cockpit lighting which was often far from ideal, this instrument was fatally easy to misread. The high work-load associated with rapid descent in a crowded terminal area makes it all too easy for the pilots to lose touch with the altitude situation when the altimeter is unwinding at nearly 100 ft/second.

After a spate of accidents in the mid-1950s in which pilots had flown into the ground through misreading their altimeters by 10 000 ft, it became generally acknowledged that the presentation of tens of thousands of feet left much to be desired. A 'ten-thousand-foot warning light' was therefore installed on the instrument panel adjacent to the master altimeter. This flashing green light was activated by an aneroid capsule whenever the aircraft passed through a pressure level equivalent to a height of 10 000 ft. The light continued to flash until cancelled by the pilot – a positive action which served to impress the event upon his mind – and the device has proved successful in preventing this particular type of pilot error.

Mis-setting errors

Unfortunately other accidents attributable to the altimeter still occur, though these are more often caused by mis-*setting* the instrument rather than mis*reading* it. The altimeter is merely an extremely accurate aneroid barometer, with its scale calibrated in units of height instead of pressure, and with a wide range of adjustment to the basic setting, which is controllable by the pilot. In order to translate the pressure reading into accurate information about height, this basic setting must be changed periodically.

Possibly the most crucial change occurs during the initial descent towards the airport of the intended landing. At a certain point during the descent the altimeter must be changed from a uniform, internationally agreed standard setting (which has been used to ensure vertical separation from other aircraft en route), to the local setting, which will give an accurate height above ground (or sea) level.

A classic example of a genuine pilot-error accident due to mis-setting the altimeter occurred in February 1964, near Nairobi in Kenya. The First Officer was making a night approach to this high-altitude airfield, and in order to ensure that his altimeter would register height above the runway, he had requested the aerodrome setting from Nairobi Control. This was passed to him as 839 millibars, 'a long wind-down' on the subscale from the standard setting of 1013 mbar. Preoccupied with the task of flying the aircraft, he mentally transposed the first and last figures, and stopped winding down when the subscale showed *938* mbar – which, as he explained later, gave him 'a visual appreciation of 839'. Since each millibar on the subscale is equivalent to some 30 ft on the height scale, his altimeter was now in error by 3000 ft.

Following normal procedure, the Captain had set *his* altimeter to the sea-level pressure, so that the difference between the readings of the two altimeters would equal the height of the aerodrome – some 5300 ft. He should then have compared his reading with that of the First Officer's instrument, to ensure that no error had occurred during the resetting procedures. This he had omitted to do by the time the aircraft struck the ground some 10 miles short of the runway, while the First Officer's altimeter was reading 3000 ft.

Fortunately the undercarriage was already lowered, and the aircraft touched down on flat open ground in the Nairobi National Park, sustaining only minor structural damage and enabling a safe landing to be made at Embakasi airport. As a result of the enquiry into this accident, the airline, BOAC, fitted their aircraft with a third, stand-by altimeter and tightened up the operational procedures for cross-checking the altimeters.

New instruments: familiar errors

The authorities in the UK meanwhile had set up a committee to investigate all aspects of altimeter problems. The report of the UK Altimeter Committee, published in 1965, recommended that the three-pointer altimeter should be replaced by a single-pointer type with a digital counter. Recent developments in the technical field had now made the electrically-driven servo-altimeter a practical proposition and it was no longer necessary for the barometric capsules to provide all the power to drive the complicated system of gearing to the pointer and drums. Power was now provided by a miniature electric motor, controlled by a system of detectors which measured the expansion or contraction of the capsules.

In the new altimeter, the single-pointer records hundreds of feet as before, but the tens of thousands, thousands, hundreds (and even 50 ft units on some types) are displayed on a digital counter inset in the dial in a manner similar to the mileage recorder on a car speedometer.

Although all the major airlines gradually adopted the new altimeter, as well as introducing more stringent procedures for cross-checking the readings, accidents and incidents attributable to the altimeter still occur – usually when the pilot's concentration on the complicated routine of the descent check list is interrupted by some non-standard procedure, or is impaired due to fatigue.

Five and a half years after the Nairobi incident, in November 1969, the first-ever fatal accident to a VC 10 occurred at Lagos during a direct approach over low-lying terrain to Runway 19. The aircraft flew straight into the trees some 8 miles north of the airport, and all on board were killed. The official report of the accident stated: '... The aircraft's descent below the Obstacle Clearance Level may have been due to insufficient attention being paid to the altimeter indications by the crew because of their preoccupation with other matters ... In addition, they were coming to the end of a long overnight flight involving three sectors and short-term fatigue could well have been a contributory factor.'

As recently as 1972, BALPA received a confidential incident report, describing how a complete crew omitted to make the change-over from standard to local altimeter setting during an instrument let-down at Mahé in the Seychelles. The approach was being made during a thunderstorm at the end of a $12\frac{1}{2}$ hour night duty period, when the pilots had been awake for $17\frac{1}{2}$ hours.

Since the normal static-free VOR (Very high frequency Omni-Range) was out of commission, the let-down was being made on the stand-by Medium Frequency Non-Directional Beacon—a much less accurate and reliable navigational aid, particularly during a thunderstorm, when the radio-compass needle, instead of pointing to the beacon, spends 50 per cent of its time pointing to the latest lightning discharge.

The Captain was concentrating on flying the aircraft as accurately as possible in the turbulent conditions prevailing, while both co-pilots were so busy timing the let-down procedure and attempting to cross-check with the Captain on the probable position of the aircraft according to the random indications of the wildly swinging ADF (automatic direction-finding equipment) needle, that the entire crew forgot to change the basic altimeter settings half-way through the let-down. The result was a dangerously small ground clearance of 500 ft during the latter stages of the descent through cloud. Had the local barometric pressure been lower (as happens during the monsoon period), the aircraft would probably have crashed into the hills.

Yet another accident due to 'pilot error' would have been added to an already impressive list; but it is interesting to speculate what degree of blame (if any) would (or should) have been attributed to an airport authority which failed to maintain its only static-free let-down aid in service at a time of year when local thunderstorms are of frequent occurrence. (For description of a similar incident see Chapter 5, pp. 193–4.)

Solutions and costs

During 1972, the National Transportation Safety Board of America became so alarmed by the continuing number of aircraft which flew into the ground during the approach to land, that it promulgated an operational requirement for a terrain proximity warning system to be fitted to civil transport aircraft. This would probably be developed from the so-called radio altimeter, which gives an instantaneous presentation of the height of an aircraft over the land or sea directly beneath it. The new system would give warning of terrain ahead and to the sides of the aircraft, but was expected to require a 10-year development programme before it would be acceptable to airline operators. (In 1974, however, a report named the first US operator to install terrain warning. Pan American announced a decision to spend a reported £1 300 000 on terrain proximity equipment for installation on their entire fleet of 140 jet aircraft (*Daily Telegraph*, 10 May 1974).)

A solution to the particular problem of remembering to reset the altimeters during descent is available now – at a price. For the past 8 years, some US military jet transports have been fitted with an improved form of vertical display altitude indicator, in which the altitude scale is printed on a moving tape, which runs between two reels, rather like a typewriter ribbon. Some 6–8 inches of this tape is visible behind a vertical window on the instrument panel, and the whole tape can be several feet in length, using a logarithmic scale to expand the presentation for the lower altitudes. The tape mechanism is controlled from a central air-data computer, and it would be a simple matter to programme this computer with the transition level for the airport of destination before take-off. Then, if the aircraft subsequently descended through this level without the pilots having reset their altimeters, a warning system would be activated.

Staying on the Runway

The operation of a modern civil airliner on world-wide routes through countries with widely divergent standards of wealth and technical development is an extremely complicated procedure. The responsibility for attempting to bring order out of chaos, and for ensuring the maintenance of a minimum operational standard on a world-wide basis is, therefore, vested in the International Civil Aviation Organisation, more usually known as ICAO.

But like its parent body, the United Nations, ICAO has no teeth. It cannot compel even those states which are full members to implement all its innumerable standards and recommended practices, or ensure that each particular department in all its member states is kept fully briefed on the operational requirements of the others. And nowhere is this lack of liaison more apparent to pilots than in the implementation of Part III of Annex 14 of ICAO – that part of the documentation dealing with 'the physical characteristics of runways'.

Ideally, a runway should be as flat and as level as drainage requirements permit, and the surface should provide a good coefficient of friction – to provide safe braking – in both wet and dry conditions. In practice, runways around the world, when viewed in side elevation, vary from convex (humped in the middle), concave (slumped in the middle) to undulating (i.e. uphill/downhill). A 10000 ft runway may even have 100 ft difference in elevation between its ends. Reference has been made to the many locations at which the runway surface is so uneven that it is quite impossible to read the essential instruments (during the latter stages of take-off) because of the vibrations transmitted by the hammering nose-wheel; while at even more airfields, after the first shower of rain, the runway surface bears such a close resemblance to a skid-pan that the fact is officially recorded in the *Pilots' Route Book*.

The Aerodrome Plan for far too many airports carries the warning: 'Caution, runway ... slippery when wet', or worse: 'Caution, runway ... extremely slippery when wet.'

Such shortcomings are perhaps understandable at those airports sited in states which only comparatively recently have become involved in international aviation – often those of the 'emergent nations'. But runways which are slippery – or extremely slippery – when wet, exist even in the richest and most sophisticated countries.

To quote an outstanding example, Kai Tak International Airport at Hong Kong was constructed some 20 years ago as a single runway on a narrow artificial peninsula running out into Kowloon Bay. The area is subject to sudden heavy showers and to strong gusty winds, and since, in addition, one end and both sides of this runway are surrounded by deep water, pilots think it highly necessary that the runway should be safe to use under all weather conditions. Yet as recently as July 1973, the *IFALPA News Bulletin* contained the following item:

> Concerning Hong Kong Airport, a number of Deficiency and Incident reports have been submitted by pilots who have had bad experiences during landing due to the coefficient of friction being virtually nil when [the runway is] wet.

The *Bulletin* goes on to describe how various aircraft have terminated their landing runs by sliding out of control across the runway with cross-wind components as low as 10 knots. One DC8 landed towards the city of Kowloon in wet weather, and only just managed to stop before running off the end of the runway (into a busy traffic roundabout), with all engines still in reverse, but with the aircraft slewed 90° across the runway end.

At this, and similar airports, all that the airline operator can do is to reduce the maximum take-off and landing cross-wind component (normally 25–30 knots) to what he hopes is a reasonable compromise between safety and regularity for wet-runway conditions, and include this information in his *Operations Manual*. All that the pilot can do when faced with a landing on a slippery wet runway is to observe these lower cross-wind minima, attempt to control his threshold speed even more accurately than usual, or divert!

The runway in winter: braking problems

The problem of wet and slippery runways pales into insignificance, however, when compared with those produced every winter by slush, snow and ice on the runway. Because it is a transient state, slush conditions do not occur very often, nor are they of long duration. This is just as well, since it is extremely difficult to measure the average depth of slush on a 10 000 ft runway, and it is the average depth which is so critical for take-off. Slush has a particular retarding effect on take-off (see the account of the Ambassador accident at Munich, pp. 49–53) and –in great part thanks to that Captain's fight against a pilot-error verdict –*every aircraft Flight Manual now quotes a depth of slush beyond which take-off must not be attempted*. It would be of interest to know when this information was first included in the Flight Manuals used by German airlines. For the relevance of this comment see pp. 50–3 and Chapter 6, p. 248.

The pilot's problem begins when the reported slush depth is less than this figure. How accurate are the measurements? How recently were they made? And how unstable are the conditions on the runway surface?

He knows that if he does attempt a take-off and acceleration appears to be subnormal, he must make his decision to abort well before V_1, since the braking qualities of a slush-covered runway are far from ideal; and he is equally aware that the assessment of these qualities is by no means an exact science.

Snow is normally a major problem only at those airports where it is of rare occurrence, so that neither the sophisticated and expensive equipment needed to clear the huge areas involved, nor the trained staff to operate it, are likely to be available. Since one individual item of mechanised snow-clearance equipment can cost from £20 000 to £50 000, it is obviously difficult to persuade any airport authority to lock up several hundreds of thousands of pounds in capital equipment which will spend most of its life taking up much-needed space, rusting away in a corner of a hangar, to be used for only 3 or 4 days every other year. Yet nothing paralyses a poorly equipped airport more quickly than a sudden, heavy snowfall.

Major airports in Scandinavia, Canada and the eastern seaboard of the USA, where heavy snow is a feature of the landscape for several months every year, are, of course, well equipped with the most modern equipment and well-trained staff to cope with snow clearance. But however well the runways have been ploughed, swept and sanded, there will still be patches of snow and ice remaining; and obviously the surface will not provide equivalent braking action when compared to its clean, dry state.

The braking action is then subjected to an approximate measuring technique, by various ground vehicle devices such as the Mu-meter, Tapleymeter, and Skiddometer, etc. The value of the braking coefficient, or 'Mu' value, is then usually translated into a descriptive adjective, and the pilot is informed: 'Braking action runway... poor' (or 'good', 'medium' or even 'poor to medium' or 'medium to good').

The system has many disadvantages, and most pilots consider that no piece of equipment weighing some 1000 lb and moving at up to 60 knots can really simulate the braking effect on a 250 000 lb aeroplane moving at 150 knots. However, the readings obtained can still give the pilot a rough guide to the sort of conditions he will encounter, provided the same basic assumptions are always used in the calculations.

Unfortunately, this has not always been the case; during the winter of 1972–3, the same Mu-meter reading meant different things to the Airport Authorities on either side of the Atlantic. In the UK, a Mu-meter reading of 0·50 or more was 'good', 0·49–0·40 was 'medium'; below 0·40 was 'poor'. The United States authorities were far more optimistic: 0·40 or more was 'good', 0·39 was 'medium', or 'medium to good'; below 0·39 was 'poor'. A pilot on the Atlantic route had therefore to remember that 'good' in the US was only just inside the 'medium' classification by UK standards – a highly dangerous state of affairs, extremely conducive to pilot error. In the past, IFALPA has made strong representations on this subject to ICAO in an endeavour to ensure that, where such discrepancies exist, international standardisation of these classifications can be achieved.

Ice on the runway, provided that it is reasonably smooth, can be sanded to provide an operationally acceptable surface for a short period. The sand is soon blown away by jet efflux, and the whole process then has to be repeated, causing the runway to be temporarily closed to traffic. For this reason, some airport authorities are loath to re-sand as often as most pilots consider necessary, although ridged ice, too (caused by the re-freezing of slush), can be yet another source of the nose-wheel vibrations which make it difficult and often impossible to read essential instruments during the later stages of take-off.

Training accidents

Accidents which occur during crew training do not directly affect the travelling public; yet they have been so numerous during the past 15 years that they form a considerable percentage of the total of pilot-error accidents, and so must merit discussion.

It is worthy of note that nearly all these accidents occurred when practising manœuvres with one or more engines inoperative, and that, in the 10-year period 1959–69, no less than 14 accidents were due to loss of directional control by the pilot when practising engine-out manœuvres. It is of equal significance that all of these accidents occurred to large American 4-jet transports, fitted with podded engines slung under the wing, well outboard of the fuselage; namely the Boeing 707, DC 8, and Convair 880.

To date (1976) British designers have never mounted jet engines in pods under the wing – they have either buried them in the wing roots (Comet), or mounted them at the rear of the fuselage as in the VC10 and Trident aircraft. In all cases engines are located close to the centre-line of the fuselage. The yawing moment of asymmetric thrust when the outboard engine is mounted 50 ft from the aircraft centre-line is obvious; a practice landing made with two engines out on the same side means that the aircraft is in an extremely critical condition during the latter stages of the approach.

The rules which govern aircraft certification and pilot training for engine-out conditions were developed for piston-engined planes, and *have never been changed*, although jet engines in service have proved far more reliable than their predecessors. The requirements insist that any 4-engined aircraft must be able to continue a take-off from V_1 on 3 engines, must be able to overshoot from a balked landing on 3 engines, and must be able to continue a circuit and land with 2 engines out on one side. So before a Captain who is converting onto a 4-engined transport aircraft is checked out, he must demonstrate his ability to perform all these manœuvres, and then prove his continuing competency at each 6-monthly check for the rest of his career.

It follows, therefore, in the light of the critical performance of the swept-wing jet with podded engines in the engine(s)-out configuration, that the jet pilot is exposed to a very real risk during his proficiency checks; as the accident record testifies. Ever since the first fatal accident to a Boeing 707 during engine-out training in February 1969, the American Air Line Pilots Association (ALPA) has consistently opposed the requirement to demonstrate pilot proficiency in these manœuvres in the aircraft, particularly at low altitude, and using the very low speed margins achieved by expert test pilots during certification.

In the ALPA paper 'The case against engine-out flight training'* it is noted that, during the 40-month period ending in December 1969, there were 51 jet transport accidents resulting in deaths and/or the total destruction of the aircraft. Of these, 9 aircraft were destroyed and 26 crew-members were killed during engine-out training. Over a 10-year period: 'Of total world-wide air carrier flight hours by jet transport, only some 2 per cent has been devoted to flight training. Yet during this 2%, approximately 17 per cent of the fatal accidents occurred.'

The paper goes on to state: 'It is significant to note that during the 40-month period referred to, not one accident occurred on the line resulting from an engine-out emergency', and quotes in some detail from the report of the first American pilot to carry out an emergency landing on the line with two engines failed on one side – and a full load of passengers! This pilot freely admitted that he made his approach in what he considered to be 'a common-sense way' – that is, at a considerably higher speed than that recommended in the Flight Manual, and that he delayed extending flaps and undercarriage until a mere 3 miles from touchdown. The ALPA paper then comments: 'He would have failed a check ride miserably with the procedure he used for the actual event, but that only serves to point out that dangerous, slow-speed check requirements have no relevance to line pilot common sense.'

In spite of this pilot's successful landing, training procedures in the USA continued to insist that the approach speed with 2 engines out on one side must be allowed to decay to 25 knots below the minimum control speed for this configuration by the time the threshold was reached. This meant that if the pilot made the slightest misjudgement towards undershooting during his final approach, there was no possible way of preventing an accident. Once the speed had dropped below the

*T. G. Foxworth and H. F. Marthinsen (ALPA): Paper 71–793, American Institute of Aeronautics and Astronautics, Aircraft Design and Operation, Meeting 3, Seattle, Wa., 12–14 July 1971.

minimum control speed, any increase of power on the two remaining engines would merely result in the pilot's losing directional control of the aircraft, since there would be insufficient airflow over the control surfaces to render them effective.

Small wonder, then, that ALPA has continually pressed for engine-out training to be conducted at altitude (where surplus height can be traded for airspeed if a dangerous situation develops) or in the simulator. Unfortunately, the FAA has been unwilling to permit many airlines to conduct engine-out training in the simulator, since several older models of these systems did not reproduce the aircraft's engine-out performance with sufficient accuracy. Re-design of such simulators would cost the airlines money; and even more expensive would be the lack of training facilities while the necessary modifications were being incorporated.

The problem is illustrated in the 1972 NTSB report on a fatal engine-out training accident to a Boeing of Western Airlines, wherein it was stated that the airline's B-720 B Flight Simulator 'did not properly simulate aircraft performance under conditions of asymmetric thrust, in that the effects of sideslip–roll coupling were easily countered (on the simulator) with the excess lateral control that was available'. The conflict here between the introduction of realistic simulation at high cost, and the alternative of in-flight engine-out training at a potentially much higher price, requires no further comment. (But see Chapter 6, Doctrine of the reasonable man, p. 208.)

The Instant Judgement–I

Earlier, this chapter asserted the potential for conflict between pilot-error verdicts and true accountability. It follows that where such a dubious verdict remains in being, a major injustice is perpetrated against the pilot concerned.

True accountability, then, must be a matter for the patient and time-consuming establishment of fact; for the questioning of witnesses, the collation and analysis of evidence, and the assessment of the results of all necessary and possible technical enquiry. It is clear, therefore, that no judgement which by default leaves any vital submission unexplored can be regarded as being in any way valid. And it must be equally plain that no assumption, or statement made within a few hours, or even a few days, of an air accident should possibly serve as a pivot for the formal enquiry. Yet the extent to which both of these elements may have contributed to the professional degradation of one pilot, and to the extinction of the career of another, should be judged from the following accounts.

The first pilot to suffer in this way was Captain R. E. Foote of BOAC, who was in command of a Comet I when the world's first jet-liner suffered its first serious accident after being in service for less than six months. Taking off from Ciampino Airport, Rome, on a dark and rainy night towards the end of October 1952, he found that the aircraft was not accelerating as it should, and in consequence, abandoned the take-off. The aircraft overran the runway and was severely damaged, though the total injury to passengers and crew was one cut finger.

The 'instant' verdict in this case was announced in the afternoon of the very next day, and appeared in the press the following morning. The conclusions are repeated here, to illustrate not only the chosen theme, but also the manner of its very considerable embellishment.

COMET ENGINE 'LOST POWER' – BOAC STATEMENT ON ROME CRASH . . .
. . . A BOAC spokesman said in London yesterday afternoon that available evidence suggested that the accident was due to one of the port engines losing power during take-off. This caused the plane to swing as it became airborne.

The pilot, with great skill, made an immediate landing on soft ground beyond the end of the runway . . .

It is generally recognised that the most critical period at which an engine failure can occur is after a plane has just left the ground and when the undercarriage is still down or partly retracted, and normal climbing speed has not been reached.

. . . Jet-engined planes are particularly vulnerable at this time. This is because, as stated in the article on the Rolls-Royce Conway by-pass jet engine in the *Daily Telegraph* on Thursday, simple jet engines accelerate slowly and do not develop full power until a plane is travelling at high speed and a considerable volume of air is being forced through the engines . . .

The report goes on to discuss methods of overcoming this 'critical period' by fitting '. . . either jet or piston-engined planes with rockets to assist take-off', and – this time without reservation – supports the engine-failure assertion in its caption to the adjacent picture of the damaged aircraft:

. . . The Comet airliner lying on the soft ground beyond the end of the runway . . . after making a forced landing on Sunday. The mishap was due to a port engine losing power as the plane became airborne.

(*Daily Telegraph and Morning Post*, 28 October 1952)

This 'off-the-cuff' diagnosis not unnaturally incensed the manufacturers (of both engines and airframe), who showed no mercy on the unhappy Captain Foote when the subsequent investigation proved that the engines had behaved perfectly. Captain Foote was to blame for the failure to take off – he had inadvertently lifted the nose of the Comet too high during the later stages of the take-off run and the greatly increased drag in this nose-high attitude had prevented the aircraft from accelerating normally. It was pointed out that the relevant paragraph of the Comet *Flight Manual* clearly stated: 'At 80 knots the nose should be raised until the rumble of the nose-wheel ceases. Care should be taken not to overdo this and adopt an exaggerated tail-down attitude with consequent poor acceleration.'

Taking off to the south, towards the unlit bulk of the Alban Hills, on a dark, wet night, Captain Foote had no visible horizon to guide him – and, because the controls on the Comet were hydraulically operated, he had been deprived of the 'feel' which had been so helpful to all pilots on earlier, slower aircraft. In the majority of these earlier aircraft, the pilots' controls were directly linked to the control surfaces; they could thus have a direct indication of the pressures imposed on these surfaces by the airstream. An experienced pilot could interpret the varying pressures he had to overcome in order to move the controls, and adjust his movements accordingly. On the Comet I the control surfaces were moved by pistons actuated by hydraulic fluid under extremely high pressure, and movement of the flying controls in the cockpit merely opened or shut the requisite valves controlling the flow of high-pressure fluid.

In the admittedly difficult circumstances pertaining at the time of Captain Foote's take-off, he should apparently have relied on a 'sixth sense' to prevent the error of 2–3° excess incidence which in fact occurred. He was informed that there was no longer a place for him amongst the élite pilots of the Comet team, though he was allowed to retain his command – demoted to the lowliest task for a BOAC pilot – flying York freighters. The York was a hurried post-war conversion from the Lancaster piston-engined bomber: out-of-date, comparatively slow, and unpressurised (and thus hot and uncomfortable for its pilots on the tropical routes they flew).

Less than 5 months later, another Comet crashed in exactly the same circumstances as Captain Foote's. During a night take-off from Karachi,while on a delivery flight to Australia, the Comet IA overran the runway, and burst into flames. The crew of five, and six technicians (the only passengers) were all killed. The manufacturers thereafter carried out a series of tests on the take-off characteristics of the Comet I wing, and as a result the wing design was modified, making it much less critical to small errors of judgement in estimating the exact angle to which the nose of the aircraft had to be raised during take-off.

That such a modification was considered necessary by the manufacturer strengthened the already widely held view among Captain Foote's fellow pilots that he had been unjustly penalised; firstly by the harsh finding of the enquiry, and then by his subsequent treatment by BOAC. BALPA attempted to have the finding modified but, after a lengthy correspondence, the final letter from the Chief Inspector of Accidents, written in May 1954, was to the effect that he saw no reason to alter the original finding.

Fortunately for Captain Foote, the International Federation of Airline Pilots' Associations (IFALPA) had become interested in the case. That organisation's Technical Secretary, Captain C. C. Jackson (himself an ex-BOAC pilot) refused to accept the decision, and carried on the battle with the manufacturers.

The unjustified slur cast upon their engines, however, had done nothing towards the creation of a receptive climate for other submissions and it proved to be extremely difficult to obtain any concession from them that their first wing design for the Comet I may have been in any way suspect. It was not, in fact, until 1959 that IFALPA, after a sustained challenge of the reduced take-off margin above the stall which had first been applied to the Comet I, managed to obtain the manufacturer's consent to publish a somewhat guarded 'agreed statement' which at the least did something to diminish the damage to Captain Foote's professional reputation. And it is to be noted that this battle to obtain even partial satisfaction had taken more than 6 years.

It is ironic that in the interim Captain Foote's courage and skill had long since retrieved his personal status in his own airline. On 31 August 1955, a propeller disintegrated while he was flying his York freighter across the Bay of Bengal. Fragments of this propeller hit the adjacent powerplant, and this caused a second propeller to break up. Swept back in the airstream, flying debris damaged not only the starboard wing, but also the starboard elevator. With only the two port engines functioning, Captain Foote flew the crippled York 100 miles to Mingaladon airport in Burma, where he made a successful two engine-out landing in marginal weather conditions with only partial elevator control.

Very properly, he was officially commended for this outstanding feat of airmanship, and shortly afterwards his application to join the newly-formed Britannia Fleet was accepted.

The Instant Judgement – II

The next pilot whose ordeal was to begin with an unguarded and premature post-accident statement was Captain James Thain of BEA.

During the late afternoon of 6 February 1958, Captain Thain attempted to take off in his Airspeed Ambassador – an aircraft popularly called the 'Elizabethan' – from a slush-covered runway at Munich, in Germany. The aircraft failed to accelerate normally and, after the pilot had abandoned the take-off at a rather late stage, he was unable to stop within the confines of the runway. The aircraft careered off the end at considerable speed, veered to one side, struck a house and some trees, broke up and caught fire. Amongst the 23 people killed and numerous injured were members of the famous Manchester United Football Club, returning from a match in Belgrade; inevitably the accident received more than the normal amount of publicity.

The Chief Executive of BEA, Mr Anthony Milward, flew to Munich the next day, and spoke to the press on his return to England on 8 February. He is recorded* as having said: 'I cannot give the cause of the disaster, but what would have been a simple mishap in which people might have climbed out of the aircraft with ankle injuries was turned into a major disaster by the house situated 300 yards from the end of the 2000 yards Munich airport runway.

'It was not an ordinary overshoot by the Elizabethan; I call it an extraordinary overshoot – failure to take off. Let us say the house should not have been there.'

The house, however, proved to be just outside the boundary of the 'cleared strip' surrounding the runway, and the size of the strip conformed to international safety regulations for airport design. As the Germans had no difficulty in demonstrating at the subsequent enquiry, the house was contravening no safety law in occupying the position that it did. They, in their turn, manifestly rejected the implication that they had tolerated a take-off hazard at their airport at Munich; in the same

* Reported in *The Times*, 10 February 1958.

newspaper report, the West German Traffic and Transport Ministry had countered Mr Milward's statement by announcing that the 'probable' reason for the aircraft's failure to take off was 'ice on the wings'.* Thus, several months *before* the official enquiry into the accident, the Germans (who would be responsible for conducting it) had publicly committed themselves to a statement of the conclusions they expected to reach.†

That such a finding was subsequently recorded can only be described as a tragedy; for the ensuing contention was not only to cost Captain Thain some ten years of unremitting struggle to clear his name, but was also, in the opinion of his fellow pilots, finally to destroy him.

The Commission of Enquiry set up by the German Ministry of Transport published its report in January 1959, almost a year after the accident. It gave as 'the decisive cause of the accident' the fact that a layer of ice had formed on the top surface of the wing while the aircraft was on the ground at Munich. This layer of ice was assumed to have 'considerably impaired the aerodynamic efficiency of the aircraft', and so prevented it from accelerating to the required speed for take-off within the runway distance available. Although evidence as to the actual existence of this ice was, at best, conflicting rather than definite, Captain Thain was held to be responsible for the accident, because he did not physically check to see that the wing was clear of ice before attempting to take-off.

---

* Reported in *The Times*, 10 February 1958.

† These passages are also based on pp. 49–50 of S. Williamson, *The Munich Air Disaster*, Cassirer, London, 1972. However, objective reporting demands that further significant statements made at that time should be noted here (Ed.).

Mr Milward also observed, according to *The Times*, that 'the effect of the weather was a distinct possibility as the cause of the disaster. "We shall have to investigate it very carefully," he said. "The captain came back after two runs and told the BEA engineer he was not happy. When he went out the third time he was entirely satisfied."'

Mr Milward said there was no de-icing spraying at Munich because the aircraft was in transit, and soft unfreezing snow was falling.

The Times report concluded with the following, under the heading 'Lift reduced': 'Our Aeronautical Correspondent writes: "The effect of snow on an aircraft's wings is to reduce lift. Snow on a runway might also retard an aircraft's speed during take-off. If ice forms on a wing surface it can be sprayed with a de-icing fluid, but soft unfreezing snow can be expected to be blown off during taxi-ing and the take-off run."'

It is striking that this correspondent's comment on the effects of 'snow on a runway' was made at this short remove from the date of the accident. See the finding of the second Fay Commission, March 1969 (this chapter, p. 52), in which this cause is confirmed. Readers should also compare the amplification of the reference to the (wholly unsupported) theory of 'wing-icing' with the report of the Comet crash, pp. 46–7. Pilots will note this syndrome with particular concern.

The report only made passing reference to the presence of slush on the runway: 'It is not out of the question that, in the final phase of the take-off process, further causes may also have had an effect on the accident.' In a later paragraph, the completely unfounded statement was made: 'All experience goes to show, however, that it may be assumed that take-offs can be made with nose-wheel aircraft without danger up to a slush depth of at least 5 cm.'

Since little was known at the time of the retarding effect of slush on the runway (but see pp. 104, 105 and Fig. 5) both Captain Thain and his fellow-pilots in BALPA considered that this assumption prejudiced the impartial examination of the causes of the accident, and efforts were made to have the enquiry reopened. This the German Commission declined to do.

In an attempt to clear his name and thereby return to the duties to which BEA had not yet restored him, Captain Thain asked for, and obtained, a British Court of Enquiry. Set up in April 1960, under the chairmanship of Mr Edgar Fay, QC, the enquiry was given – in the opinion of pilots – extremely limited terms of reference in order to avoid political embarrassments with Germany. When the Fay Commission Report was published in August 1960, however, Captain Thain was not absolved of all blame, since it was found that he did not '...take sufficient steps to satisfy himself that the wings of the aircraft were clear of ice and snow'. And although the Fay Commission did not agree with the German assumption that 'take-offs can be made...without danger up to a slush depth of at least 5 cm', they still appeared to agree with the original finding that the major cause of the accident was wing icing.

First admissions

It was less than six months later, in January 1961, that the British Ministry of Aviation published an *Information Circular* on the effects of slush on take-off performance. In direct contradiction to the confident German assumption that '...take-offs can be made with nose-wheel aircraft without danger up to a slush depth of at least 5 cm', the *Circular* warned pilots that only half an inch of slush on the runway would increase the take-off run by 40 per cent for nose-wheel aircraft ($\frac{1}{2}$ inch = 1·27 cm – about one quarter of the 'safe depth' quoted by the German Commission).

Meanwhile IFALPA had taken up Captain Thain's case, and in 1962 the Federation asked the German Commission to reopen their enquiry, since 'new evidence' (the continuing investigation both in the USA and

UK into the effects of slush drag) had become available. In a written decision made public in January 1963, however, the German authorities gave their detailed reasons for refusing this request – none of which was acceptable either to IFALPA or to Captain Thain.

Nearly 5 years had now elapsed since the accident had occurred; and during that time Captain Thain had never flown again.

In July 1964, the British Ministry forwarded to the Germans yet more details of further slush drag experiments – this time carried out on an Ambassador aircraft identical to Captain Thain's. After deliberating for another year, the Germans finally agreed to reopen their own enquiry in November 1965. The report was not published until August 1966, when it still maintained that 'Wing icing was an essential cause of the accident.' Somewhat grudgingly, it added 'Slush ... was a further cause.' It appeared that in spite of the very considerable increase in technical knowledge about the effects of slush drag in the intervening $8\frac{1}{2}$ years,* there was as yet insufficient weight of evidence to make the Germans publicly reverse their original, although increasingly insupportable, decision.

Neither Captain Thain nor BALPA gave up, however; and finally, in April 1968, the Board of Trade (which had now taken over responsibility for civil aviation) was persuaded to reopen the 1960 Court of Enquiry – this time with wider terms of reference. The Chairman was once again Mr Edgar Fay, QC, and the enquiry therefore became known as the 'Second Fay Commission'. After considering the very substantial mass of new evidence available, the Commission was instructed 'to report whether, in their opinion, blame for the accident is to be imputed to Captain Thain'. The report of the Second Fay Commission† was presented to the President of the Board of Trade in March 1969, and made public in June of that year.

The report completely reversed the original finding of the German enquiry, stating that 'ice (on the wing) may possibly have contributed to the accident, but we think it unlikely to have done so ... In these circumstances we cannot and do not find blame.' The report went on to state that, in the light of present knowledge, *it was their opinion that slush on the runway was the prime cause of the accident. However, so little had*

---

* References to investigations into slush-drag effect by American Airlines, KLM, NASA, the FAA, Trans-Canada Airlines and Britain's Royal Aircraft Establishment, appear in *The Munich Air Disaster* (see pp. 61–2, 115–16, 202–3, 205–6, 213–14).

† CAP 318, *Report of the Second Independent Review appointed to consider the accident to Elizabethan aircraft G-ALZU at Munich on February 6th 1958, and to report whether blame is to be imputed to Captain Thain.* 18 March 1969.

been made public on this phenomenon 11 years ago that Captain Thain could not have been expected to take this into account. Finally: '*In accordance with our terms of reference we therefore report that in our opinion blame for the accident is not to be imputed to Captain Thain*'.

It had taken over ten years of almost continuous battling against officialdom in two countries for one indomitable pilot, ably sustained and abetted by both BALPA and IFALPA at various stages, to reverse an unjust verdict. During these ten years, Captain Thain had been dismissed by BEA (although compensated for loss of earnings); he had been submitted to the mental strain of being held responsible for the deaths of 23 people; and he had witnessed the futile ending of a career of professional service.

In the event, he was to be denied, too, the full moral restitution which should have followed from his vindication, for the German authorities – clearly unimpressed by the British (and now world-wide) acknowledgement of the true cause of the Munich air disaster – have persistently refused to accept the findings of the Second Fay Commission, without producing valid reasons for this refusal. (But see legal implications, Chapter 6, pp. 247–8.) If there can be an adequate epitaph to this sorry episode in aviation history, let it be the following news item of 7 August 1975:

> Captain James Thain, 53, pilot of the BEA plane which crashed at Munich in February, 1958, killing 23 people, including eight Manchester United players, has died at his Berkshire farm. Captain Thain, who died without being cleared by the German authorities of responsibility for the crash, had recently suffered from heart trouble.
>
> He had ardently sought exoneration by the Germans for 17 years. He was cleared by the British Government in 1969. Captain Thain was suspended by BEA soon after the disaster, in 1960.
>
> At the time, the British Airline Pilots' Association described his dismissal as 'a gross miscarriage of justice'.
>
> (*Daily Telegraph*, 7 August 1975)

The right of appeal

After this bleak saga it is cheering to record a gleam of light, for in the UK pilots have been provided with the opportunity to challenge the findings of the Accidents Investigation Branch before the report is made public. Thus, if a pilot feels he has been unfairly criticised, he can attempt to have the finding, or at least its form of wording, modified.

Since the United Kingdom Accidents Investigation Branch has an enviable international record of objectivity and fairness, this right is not often exercised; but the importance of its existence as a last line of defence cannot be overstated.

In fact, few pilots have the technical background and detailed knowledge to challenge an inspector of accidents on his home ground, and many – particularly among younger pilots – may be discouraged from doing so by the sheer weight of the 'official' aura surrounding the enquiry. But provided that he can convince his professional association of the justice of his cause, the pilot *can* obtain powerful support.

Most of the larger pilots' associations, through their technical committee and its associated specialist study groups, have a nucleus of 'pilot experts' in most technical aviation subjects such as airworthiness, meteorology, all-weather operations, and even accident investigation. If necessary it is always possible to apply for the service of an expert in any particular subject from another pilots' association through the good offices of IFALPA.

In the light of the comment above, the informal and sympathetic character of one such proceeding will be of interest. No less so is the unfolding development of the situation which generated the enquiry.

In September 1967, an Aer Lingus Viscount crashed (without loss of life) while attempting to land at Lulsgate Airport, Bristol, in patchy fog. This small provincial airport had none of the sophisticated 'all-weather' aids such as an instrument landing system, full approach light system, or electronic devices for accurately measuring the visibility along the actual runway in use and the cloud-base out in the approach area, all of which could be expected at a major international airport.

Moreover, by reason of its situation, it was prone to this type of fog in the early morning. The fog would lift to form very low stratus cloud as the day advanced. Under these conditions, which prevailed on the morning in question, the Captain was provided only with reports of the general visibility and cloud-base – both measured from the meteorological office a mile away from the threshold of the runway in use.

Since both the general visibility (at 1500 m, decreasing later to 1200 m), and the lowest cloud reported (at 1500 ft), were both well above his company's landing minima, the Captain accepted a Ground Controlled Approach (GCA) or 'talk-down' by radar, to position him for landing.

He saw the elementary approach light system off to his right just as he reached his 'decision height' of 260 ft, and made a corrective turn to align the aeroplane with the runway ahead. Now, according to his operations manual, the decision height is the lowest height to which the Captain may descend while making an approach on instruments. At or before reaching this height, he must 'be able to continue the approach and landing solely by visual reference'.

By inference therefore, the decision height is also the lowest height from which it is safe to make an overshoot on instruments. However, what no operations manual ever provides is guidance to the Captain as to his best course of action should he lose his 'visual reference' (by entering very low cloud or patchy fog) *after* he has passed his decision height. This is for the very good reason that there *is* no safe advice to give in this admittedly rare, but highly dangerous situation. So if the Captain does have an accident in such circumstances, he is always 'wrong' in choosing the alternative which got him into trouble.

On this occasion, the Captain elected to continue his approach and attempt a landing. He damaged his aeroplane, and so (obviously!) 'he should have attempted to overshoot'.

When the preliminary draft of the accident report was sent to him in 1968, the Captain considered that the British Accident Inspector had been unduly harsh in his findings – a view which was shared by the Irish Pilots' Association. In consequence, the author was asked by IFALPA to act as 'counsel for the defence' or 'prisoner's friend' as he is still called by ex-service pilots acquainted with court martial routine; a task which is by no means undertaken lightly, since all liaison, technical work and research connected with the case must be carried out on a voluntary basis, in the 'prisoner's friend's' own free time.

Plainly, to warrant such aid, the accused pilot must be seen to have a strong *prima facie* defence against the charge. The evidence in this case made it apparent that the Captain had been the victim of a chain of adverse circumstances, since he had had to contend with inaccurate meteorological information, a non-precision landing aid coupled with a non-standard approach light system, and an inherently dangerous combination of landing minima; the latter promulgated in all good faith by his operator, in the airline's Flight Manual. There appeared to be a sound basis for an appeal to the Inspector to reword his finding in a manner which would be much less damaging to the Captain's professional reputation.

His cause had gained him the personal support of the Chairman of his Association, and after several meetings in both Dublin and

London, a submission was entered for hearing at the London head-quarters of the Board of Trade (now the Department of Trade and Industry). The arguments put before the Inspector and his colleagues at that meeting were plainly convincing, for, at the end of a long morning during which every matter at issue had been thoroughly scrutinised, a gratified defence team heard the Inspector agree: 'I think you have made your point; do you happen to have an alternative form of wording which would be acceptable to both sides?' Since the Captain and his friends had worked into the small hours, drafting and re-drafting just such an 'alternative form of wording', copies of the proposals were immediately forthcoming – whereupon the meeting was adjourned to enable officials to consider these in private.

The suspense of the next 15 minutes – during which a professional career hung in the balance – can readily be imagined; as can the reaction to the Inspector's eventual decision: 'Your revised form of wording is quite acceptable to us.'

The final paragraph of the official report on this accident, therefore now reads: 'At a meeting on 30 August 1968 Captain D... and his two advisers made certain representations which were taken into consideration during the preparation of this report: the wording of paragraph 2.2(b) has been amended.'

Accident reports and more instant judgements

This meeting had been a most heartening experience, not only for the jeopardised pilot, but also for his two advisers; and it is worthy of note that the report was published only a year after the accident occurred. For although, as has already been remarked, pilots have the highest regard for the 'end product' of the United Kingdom Accidents Investigation Branch, it often appears to move with glacial slowness.

Taking six representative accidents over a 4-year period, the *average* time which elapsed before the report was published was 18 months, as Table 2 shows. *Note that while the shortest time was 9 months, the longest was over $2\frac{1}{2}$ years* (compare also the cases of Captains Foote and Thain, already cited).

Similarly, of the thirteen accident reports published by the United Kingdom Accidents Investigation Branch in 1975, the longest interval between accident and report was 29 months (in the case of Vanguard aircraft G-AXOP which crashed at Hochwald/Solothurn, Switzerland, in April 1973). This report was published in September 1975. The shortest interval between accident and report was 13 months – Piper

Table 2 Accident report publication dates

Date of accident	Report published	Months elapsed
May 1967	January 1969	20
November 1967	June 1970	31
April 1968	April 1969	12
July 1968	March 1969	9
March 1969	September 1970	18
June 1971	November 1972	17
	Average time 18 months	

Twin Comanche PA30 G-ATYR which crashed at Saulmore Bay, Argyll, in October 1974. This report was published in November 1975.

The average accident/report interval for the remaining eleven crashes was some 16 months.

This, of course, is in marked contrast to the regrettable tendency on the part of the mass media – and even of some authorities who should know better – to pass 'instant judgement' on the pilot, almost before the wreckage has stopped burning. In far too many cases, this off-the-cuff verdict of pilot error is returned on the flimsiest evidence.

On 13 August 1973, a Spanish-owned Caravelle crashed while making an instrument approach to the Corunna Airport in North-West Spain. The accident occurred in bad weather, some 4 miles from the airport, and all on board were killed. A report published *less than 24 hours after the accident*, stated that 'reports from Corunna indicate that the crash was due to pilot error' (*Daily Telegraph*, 14 August 1973).

Since the crash was reputed to have occurred 4 miles from the airport, taking the generally accepted rule-of-thumb approach slope of 300 ft to the mile, the aircraft should have been at some 1200 ft at the point of impact. No evidence was quoted to show how the pilot came to be 'in error' by such a large amount. As far as the dead pilot was concerned, the report conflicted sadly (and yet again) with the normally accepted principle of British justice – 'Innocent until proved guilty'.

Yet another example of an authority 'jumping the gun' occurred the day after the mid-air collision between a DC9 and a Coronado, over Nantes, France, on Monday, 5 March 1973, and once more it will be enlightening to offer the sequence of events as presented in the press reports of the disaster. Within 24 hours of the accident M. Robert Galley, the French Minister of Transport, was quoted as saying that 'the crash was caused because of instructions that were not followed' (*Daily Telegraph*, 6 March 1973). On the next day M. Galley's views were reported more fully:

Coronado turned into path of DC9, says Minister
FRENCH BLAME PILOT FOR MID-AIR CRASH
Britain's 5000 airline pilots and most world airlines yesterday banned flying over France after Monday's mid-air collision in which 68 people died ... The ban was imposed despite a categorical denial by M. Robert Galley ... that military air controllers who are doing the job of striking civilian controllers were to blame for the accident.

... M. Galley said on French radio: 'Military controllers were absolutely not the cause. A succession of pilot errors led to the incident.'

He said that the DC9 'was some minutes behind in its flight plan – probably two minutes – while the Coronado was considerably ahead on schedule. At 13.26 hours the pilot of the Coronado received an order to slow down and to pass over the beacon at Nantes at 14.00.

'He acknowledged that he had received the order twice but he did not slow down, and approximately at the Nantes beacon he made a turn to the right which put him into the way of the DC9.'

M. Galley said the controllers had acted properly but the Coronado arrived over the Nantes beacon six minutes early.

'It is possible the pilot did not understand the order but he acknowledged twice that he had received the command', he said. '(The pilot) also received an order to change radio frequency to pick up Brest control, but he did not do it.'

The Minister's statements were reinforced by a comment from General Claud Grigaud, French Chief of Staff, reported in the same newspaper:

The main cause of this accident is that the pilot of the Coronado charter plane did not execute orders from ground control for reasons we do not yet know.

The General said that the Coronado pilot had failed to take proper action in time to delay his arrival over Nantes beacon. 'He was told not to fly over Nantes before 14.00 hours,' he said.

The British newspaper report, however, carried an immediate challenge to these indictments.

Speaking for BALPA, Mr Gordon Hurley said: 'We have put an absolute ban on our pilots flying over France. It is impossible for M. Galley to know the cause of the accident in a few hours when an inquiry into such a disaster usually takes months.'

(*Daily Telegraph*, 7 March 1973)

Within a week of the Minister's accusation, preliminary investigation showed that the military controllers' methods of dealing with 'conflicting traffic' left much to be desired. The collision occurred because one controller gave an order to the pilot of the Coronado which could not possibly be carried out; while the DC 9 pilot remained in complete ignorance of any impending danger.

Both aircraft were flying at 29 000 ft along separate, though converging airways which joined at the Nantes radio beacon, and both were estimating their time of arrival over this beacon at approximately the same moment. Some 16 minutes before this (although the newspaper report indicates 34 minutes' warning), the pilot of the Coronado was told to slow down his aircraft to delay his arrival at Nantes. Since his was the faster aircraft, and was nearer to the beacon in terms of time, it appears that this order should have been given to the pilot of the DC 9. As soon as the Coronado pilot had slowed his aircraft down to its minimum safe speed, he realised that it was impossible for him to lose sufficient time in the short distance remaining. He therefore asked for permission to carry out an orbit but, for some reason which was not immediately made clear, he was unable to get clearance for this manœuvre.

Finally he decided to make the orbit on his own initiative, having no other alternative; but when he had half completed it, and was facing back in the opposite direction, he collided wing-tip to wing-tip with the DC 9, whose pilot (listening on a different frequency) had at no time been warned of the collision danger. The smaller DC 9 crashed, killing all on board, but the Coronado pilot managed to land his severely damaged aircraft, carrying 108 passengers, at Cognac, 80 miles away.

Here he was informed that he was to be held responsible for the crash – apparently because he had not properly carried out an impossible order which should never have been given to him in the first place! However, the official report, published two years later, was less specific.*

---

*The report, which finally appeared in the *Journal Official* on 1 March 1975, was prepared, by M. René Lemaire, Chief Inspector of French Civil Aviation. Blame was divided between the military air traffic controllers and the pilot of the Coronado, Captain Alvarez.

The following extract from the report was published in *The Times*, 1 March 1975:

'... the complexity of the organisation of the control, *certain discrepancies among the directives in use*, as well as the *insufficiency of the radio and radar facilities available*, constituted a source of difficulty for the efficient running of the approach routes to Nantes.'

On the other hand, an analysis of the Coronado's flight shows *'an insufficiently strict application*

The Public Enquiry

It seems that only if sufficient public disquiet is aroused to warrant a Public Enquiry is any sense of urgency imparted to the investigation procedures; but few pilots approve of Public Enquiries, because of the wide terms of reference granted to their Chairmen. Such enquiries, it is claimed, appear to become ever more lengthy and expensive, and more and more of what they contemptuously term a 'lawyers' benefit'. It is increasingly difficult, in their view, to reconcile many such theoretical proceedings with what should surely be the primary purpose of accident investigation – the prevention of future accidents, or at least the reduction of the future accident rate. For pilots, these aims are paramount. It is a fact, however, that they may not necessarily meet the insatiable need of the media for a readily identifiable scapegoat. Certainly, the Public Enquiry, as at present constituted and conducted, appears to devote inordinate effort to this end; and to indulge a legal interplay of attack and defence patently calculated to prepare the ground for the formidable financial claims which will follow.*

The Trident Enquiry, November 1972–January 1973

The Public Enquiry into the crash of the BEA Trident aircraft at Staines, Middlesex, in which all 118 persons on board were killed, was set up with commendable promptitude – opening only 5 months after the accident. In its handling and in its ultimate conclusions, however, it in no way succeeded in lessening the pilots' distrust of this procedure. The enquiry continued for 2 months in the full light of publicity, exploring – inter alia – such matters as the Captain's previous relations

---

of regulations as well as the particular instructions of control. The result was an exceptional situation which required of the pilot and control, constant attention for executing under instructions a manœuvre which was particularly difficult.'

On the military controllers at the Brest ground station, the report says that '*the attribution by control of the same flight level to the two aircraft, due to arrive over Nantes airspace at the same time*, created a source of conflict.

'The Coronado was moving out of the Mont-de-Marsan area control, in south-west France, and into that at Brest. But, the commission finds, the Brest controllers did not act according to regulations when instructing him (the Coronado pilot) ... to change his frequencies as he passed out of one zone into another. *He did not understand the instructions* because, it seems, *the military controllers had not used the right terms.*' Editor's emphasis.

See also Chapter 4, pp. 153–6 for discussion of these hazards, p. 154 fn. and pp. 214–15.

* Cf. Legal view on public enquiries (Chapter 6, pp. 247–50). See references on pp. 246–7 to the DC 10 disaster; and claims against McDonnell Douglas. See also references to Trident evidence, p. 249 and Munich crash pp. 247–8.

with his colleagues, and the meaning of and motivation for, graffiti scrawled on the navigation tables in the cockpit.

To pilots, it appeared from the evidence that, insofar as one major cause of the accident could be isolated, it was undoubtedly the premature retraction of the leading-edge droop. This was carried out by either the Captain or the First Officer at too low a speed for the aircraft to continue flying without the extra lift which the droop provided.

This particular pilot error would not necessarily have proved fatal of itself – indeed, in the course of the enquiry it was established that it had occurred on two previous occasions, from each of which the crew had been able to recover control of the aircraft. It was also established that, largely for reasons of poor communication, neither of these events had served to alert those concerned to 'the likelihood of a further, and now fatal, recurrence, and to the urgent need for remedial modifications to be put in hand. Nor, it was shown, was there, prior to the date of the crash, any demonstration (in training) of the effect of premature droop retraction.*

Like so many crashes, this one owed much to a combination of circumstances, none of which would necessarily have proved fatal in isolation. These included:

(1) The suggestion that the Captain had suffered a heart attack soon after take-off.
(2) The fact that two inexperienced co-pilots had been rostered to fly together.
(3) The necessity to adhere to a complicated 'noise abatement' flight profile, which, although officially 'safe', nevertheless reduced the performance safety margin below the optimum.
(4) The factors which permitted the retraction of the leading-edge droop during the critical phase of flight.

The official report of the accident listed five 'immediate causes' and seven 'underlying causes'. The very last 'underlying cause' was given as: '*Lack of any mechanism to prevent retraction of the droops at too low a speed after flap retraction.*'

The point has been made in this chapter that most remedial safety

* *Trident I G-ARPI*, report of public enquiry into causes and circumstances of accident near Staines, 18 June 1972, HMSO, 9 May 1973.

measures are expensive; and since airlines so often operate on a slim profit margin, the operator may be understood, if not forgiven, where he is seen to have tolerated an 'acceptable risk'.* It is on the definition of just what is an acceptable risk, however, that pilots and operators have some of their most bitter arguments, and many safety-conscious but disillusioned pilots have, perhaps unfairly, echoed Nevil Shute's bitter comment: 'Of course, operators are all for safety – just as long as it doesn't cost them money!' It is this conviction which generates the cynical accusation that an airline will 'accept' a risk, because to lessen it would put one or more particular departments over the allotted budget; while if a crash occurs, it is not the operator who foots the bill, but the insurer.

Because of the dangers of this belief (however untrue), both individual pilots' associations and IFALPA have for many years attempted to enlist the support of aviation underwriters in safety arguments – but without success. The insurance world steadfastly refuses to take sides in operational arguments on safety, agreeing with the airline managements that safety matters are the ultimate prerogative of the operator.†

For their part, the operators say: 'Pilots are a highly-trained and responsible body of professional men. We don't pay the high salaries they demand in order for them to make mistakes.' Self-evidently the statement equates salary-level not merely with capacity – which is proper – but with perfection; a rare state in human beings and not one in which the rewards are usually measured in cash.

It must be admitted that the arguments originally propounded by the major British operators against incorporating some new refinement or modification considered necessary by the pilots, have gradually become more practical. In the immediate post-war years, every 'pilot suggestion' for new or improved equipment was invariably countered by demonstrating how ruinously expensive it would be to fly with all this extra 'dead-weight' on board. Statisticians were employed to calculate the future loss of revenue due to losing 10 or 20 kg of payload on every flight for the remaining life of the aeroplane – and often arrived at a truly awesome total. In vain did the pilots retort that since the aeroplane was operating, on average, to a 60 per cent load

---

* But see designer's reasoning, Chapter 3, pp. 112–15 and 'doctrine of reasonable decision', Chapter 6, p. 208.
† Editor's note: It is a fact that members of a leading aviation insurance underwriters were approached as potential contributors to this book. All declined on the grounds stated. See also Chapter 5 for Captain Leibing's comments on the handling of 'safety matters'.

factor for the majority of its flights, this 'loss of payload' would only assume any real importance on perhaps one flight in ten!

The patent absurdity of the operators' argument eventually led them to discard it. Now (much more valid) the first question asked is: 'How many revenue flying hours will be lost while the aircraft is in the hangar, being modified?'

For the operator, these are perhaps legitimate considerations. In the public mind, however, there can only be disquiet when the 'acceptable risk' materialises as an accident cause; or when the lack of vital safety measures is exposed in the course of a post-accident investigation.*

Lessons

Dr Walter Tye, the Controller of Safety in the CAA, was no doubt referring to the Trident accident, when, during the course of the Twentieth Barnwell Memorial Lecture† which he gave to the Royal Aeronautical Society in 1973, he said: '. . . I do not subscribe to the view of some American lawyers who equate human failure to an act of negligence . . . In the ordinary sense of the word, people concerned with aircraft are not negligent. Human failures are not usually blameworthy, as they stem from insufficiency of knowledge or skill or foresight . . . The so-called crew error accident continues to predominate. I would prefer not to label these accidents as crew error. Rather I would view them as failure of design, or lack of provision of aids, or inadequate training or planning, which allowed human error to occur too readily or with too drastic consequences.'

Constrained, as he considers himself, by the maxim that 'extra safety costs extra money', yet living as he must with his knowledge of the 'consequences', where does the operator draw the line? He is in fact

---

* '. . . incidents before last June should have alerted BEA to warn pilots that the droops could, contrary to standard drill, be retracted too soon . . .' (*Daily Telegraph*, 10 May 1973). See also Trident report.

In making its recommendations, the report says: 'We recognise and applaud the fact that the parties and particularly BEA have already taken steps to implement in advance much of what follows.' (*Ibid.*)

Accepting the report wholeheartedly, Mr Philip Lawton, BEA chairman, said '[the airline] will follow its recommendations exactly, some of which [steps] have already been taken.

Firstly, the noise abatement procedure has been altered to increase the climb by maintaining more power, so now there is no need to retract the trailing edge flaps until 3000 feet is reached or 190 knots.

So long as the flaps are not retracted, a baulk remains in position to prevent any movement of the droop lever . . . All of BEA's fleet will also have an additional baulk fitted to the airspeed, preventing its movement at too slow a speed . . .' (*Ibid.*)

† Quoted by courtesy of the speaker.

reinforced in his ambivalent attitude to risk by the guidance offered in a 'policy statement' issued by the FAA some years ago:

> The acceptance of a calculated risk is an integral feature of daily life, and by no means peculiar to aviation. There is a calculated risk factor in every mode of transportation.
>
> Pursuit of the goal of absolute safety would impose an economically intolerable requirement for public expenditure ... it might be absolutely safe to fly but so expensive that no-one would.

It is interesting to note that one of the examples quoted of 'extra expenditure' for dubious results, was the following: 'Having two pilots on a flight deck is safer than one; but it also costs more.'

It is perhaps more hopeful for the security of the public, and more fruitful for the future of the aviation community, to return to the theme of Dr Tye's lecture:

> There is little hope of a dramatic invention which will revolutionise safety. Ideally we should pursue the safety improvements which have the largest cost-benefit; that is, those in which the ratio of the value of accidents avoided to the cost of prevention is highest. One obvious course is to apply efforts most vigorously in the areas in which the past record shows that accidents predominate ... I do not imagine that even with understanding one can overcome human frailty. But with understanding one stands a chance of alleviating the conditions which are conducive to error. I would not write off crew error accidents as unavoidable, but view them as accidents which, with better design of aircraft, instruments, ground aids, training or procedures are amenable to reduction. *An essential preliminary is for all concerned who have the means to make such improvements, to understand more of the behaviour of the man at the controls.*

2

Human Factors
Martin Allnutt

The search for an understanding of this topic is the province of psychologists and specialist workers, and the range of their endeavours is highlighted in the Appendix to this book. Less apparent, however, are the disciplines which have been evolved to support these studies. Certain of these approaches concern themselves with the identification of 'specific features' – i.e. with one type of accident or with one type of human failure. Study groups such as those set up by the British Airline Pilots' Association to investigate problems of pilot fatigue, or by other bodies with equally clearly defined terms of reference, are typical examples in this category – as are such contributions as the work by Kraft (1969) on visual clues during night landings.

For other researchers, the 'statistical approach' offers a possible key to pilot behaviour. The proportion and type of pilot-error accidents by day and night are tabulated. Critical decisions and critical elements of the occurrence are analysed, and a 'cluster analysis' used to reveal the factors common to a number of accidents.

This approach may probe even more deeply in order to show the frequency with which any breakdown in the pilot function may have contributed to an accident. Thus, a statistical evaluation (Wansbeek 1969) of 83 pilot-error accidents produced the following table of causes:

Improper flight technique	15
Over-confidence	16
Insufficient care	18
Other causes	34

Such categorisations have long been established as research groundlines. Pilot error has been defined, for example, in terms of failures of coordination and technique; or, as a consequence of shortcomings in the exercise of alertness and observation, or intelligence or judgement; or again, as a phenomenon influenced by personality and temperament (McFarland 1953)

Similarly, other analyses list among accident causes those which may have been induced by design factors or operational procedures, by ignorance, by deliberate acts of omission or commission, by environmental factors, or by psychological or physiological causes.

Approaches such as these are most necessary, since their findings serve to indicate the areas in which research should be concentrated and remedial action sought; but primarily these methods are concerned with reconstructing a sequence of events. The focus, in fact, is on what happened, rather than on *why* it happened.

The normal pilot

It is the need to answer the latter question which motivates the third avenue of research – the 'normal pilot' approach – adopted in the author's own field of aviation psychology and considered in some detail here. Essentially the basis of this method is summed in the quotation which heads Chapter 1, namely, the self-evident truth that all pilots are human.

This acknowledgement charges the psychologist with the task of unravelling the cognitive and psychomotor processes which determine not only the level of the pilot's professional performance, but also his response to demanding or critical situations; and it is this task which is aided, in great measure, by the fruits of particularly hard-won experience.

It is recognised, for instance, that the undoubted individuality to be found among this professional group is fortunately not reflected in the diversity of pilot errors on record; and, in fact, a significantly repetitive strain is manifest among those occurrences. This characteristic permits the classification of the major types of error under headings which are appropriate to behavioural areas; e.g. information processing, illusion, false hypothesis, habit, motivation, and stress and fatigue. The typical examination of these sectors in the following pages is therefore fundamental to that 'closer look' at the pilot which is the researcher's avowed aim.

1. *How does the pilot process information?*

A simple model of the human information-processing system is shown in Fig. 4. The model is, of course, greatly simplified, with no attempt to reproduce the actual profusion of associated filters and feedback loops. It will nevertheless suffice for the purpose of tracing an item

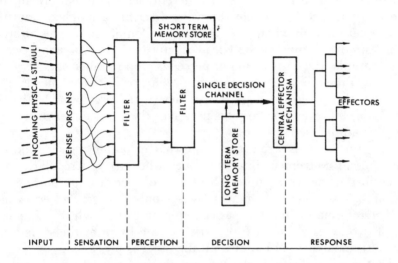

Fig. 4 Human information processing sequence

of information through the system and for some illustration of the variety of stages at which errors may occur. For example, an aircraft captain might tell his co-pilot to 'stop number one', meaning that he wants the number one engine closed down. But before the co-pilot complies with this order a few seconds later, the information must pass at least five hurdles.

(a) *Sensation.* In order to be 'heard' or 'seen' the incoming signal to the human body must register on the appropriate sense organ; however, these sense organs are both limited and, in some cases, missing. Thus, man can hear sounds only within a fairly narrow frequency band and when they are of sufficient loudness. Similarly, he has no sense of velocity, but only one of acceleration. In the above example, the message 'stop number one' was of the correct loudness and frequency for it to register at the ear, and so the first hurdle is passed.

(b) *Perception*. The fact that the auditory nerve passes a particular message to the brain is no guarantee that the brain will always deduce the same meaning from that message. What is perceived will depend on the stimulus – i.e. the action, word, or phrase used in the context in which the message is passed, and on the hearer's previous experience.

The word 'chopper', for instance, may have an entirely different meaning for, and may invoke a different behaviour pattern in, the woodcutter, who knows that the word describes a handaxe, and the helicopter pilot, to whom the word is a colloquialism for his aircraft. The stimulus, it must be noted, is exactly the same in each case. In the example, the co-pilot might perceive 'stop number one' very differently if the message were passed in a bar, or if it were the key to a joke which he and the captain had shared the previous evening. In this case, however, the co-pilot perceives the message correctly and the second hurdle is passed.

(c) *Attention*. It is a sad fact of life that incoming messages do not reach us at convenient intervals, but rather arrive irregularly, and often just at the wrong time. This fact is of special importance, since research has shown that man possesses only a *single decision channel*, and *all* information must be passed sequentially through this channel – i.e. if two items of information arrive at the brain at the same time, one of these items must wait until the other is processed.

The concept of the single decision channel is well established, of course; nevertheless, many people's first reaction to this idea is to protest that they *can* in fact do two things at the same time. Close inspection of any such demonstration shows that the person concerned is simply scanning from one source of information to another very rapidly. An obvious parallel is provided in the case of a passenger in a car being driven by a friend: at first the driver appears to be doing two things at the same time, namely driving the car and talking. But when he pulls out to overtake and finds an approaching vehicle in his path, it is noticeable how quiet he becomes. He has given up scanning and is concentrating on only one source of information.

Anecdotal evidence, too, is confirmed in the laboratory, where different messages are put into the right and left ears. It is found that when the listener attends to the input to one ear he can tell his questioner virtually nothing about the message arriving at the other ear. Nor does it matter whether the other message comes to the ear, the eye, or to the seat of the pants, that area popularly renowned as the location of a mysteriously 'instinctive' knowledge. Man, in fact, can *only attend to one thing at a time*; and it is his central decision channel which limits the speed at which he can process the information.

The function of a device such as the klaxon horn – the 'attention-getter' in the aircraft cockpit – is to ensure that a warning signal takes precedence over all other inputs and goes immediately down the decision channel. Yet there are numerous examples of warning signals being ignored, either because the pilot was dealing with more important information, or because he could not shift his attention from one input to another.

In Chapter 1, Captain Bressey refers to a number of accident findings which typify this situation, for it is with distressing frequency that 'preoccupation' is attributed as a factor in those events:

'... Preoccupation with a malfunction of the nose landing gear ... distracted the crew's attention ...' (p. 34);

'... Preoccupied with the task of flying the aircraft ... he mentally transposed the first and last figures ...' (p. 36);

'... Insufficient attention ... to altimeter indications by the crew ... because of their preoccupation with other matters ... (p. 37);

'Too busy ... timing the let-down procedure ... the entire crew forgot to change the basic altimeter setting ...' (p. 38).

It is worth dwelling on the point that, while one item of information is going through the decision channel, other items which have arrived at the same time must await their turn in, effectively, a short-term memory store. It has been shown, however, that items waiting in such a store are quite likely to be forgotten very quickly. Particularly is this true in the case of the older pilot, for laboratory experience indicates that if such a pilot is given simultaneously, say, both a new clearance and a relatively more important piece of information – perhaps the proximity of another aircraft – he may have forgotten the first piece of information by the time he has processed (i.e. considered and acted upon) the 'priority' message.

The limited capacity of man's single decision channel means that situations arise in which, even though all the component parts of the system are working well, there is still so much information that the channel becomes overloaded. The pace and stress of modern living, indeed, has made people all too familiar with situations in which they are required to attend to too many inputs. Familiar, too, are the number of well-recognised techniques which human beings adopt in dealing with this overload situation. Depending upon temperament and capacity, individuals may deal with each piece of information quickly and badly, or may concentrate entirely on one source of information to the exclusion of all others. They may confuse information obtained from two or more sources, or may even seek to escape from the situation by ignoring all the inputs, possibly by indulging in a totally irrelevant activity.

The co-pilot in our present example has many inputs which require his attention; but he decides to concentrate on the message 'stop number one' and so surmounts the third hurdle.

(d) *Decision.* Once the co-pilot hears and understands what his captain wishes him to do, the decision to comply should be, and in the vast majority of cases is, a simple one to make. He merely closes the throttle and passes onto the next decision. However, there may be occasions when the decision is not an easy one. Perhaps the co-pilot looks at his instruments and sees that it is, in fact, not the number one unit, but another engine which is malfunctioning. His decision now becomes far more complex. At a subconscious level such a situation may trigger what is in effect an internal 'pay-off' matrix, whereby the mind assesses the decision in terms of the probable outcome and its consequences. A pilot, for instance, might wish to be 'reasonably' confident that a minor instrument was functioning correctly before trusting it, but would need a far higher level of confidence in the correct functioning of a major instrument, where a malfunction might prove fatal.

Fortunately there are no complications when the message 'stop number one' is passed; therefore the co-pilot's brain sends a message to his hand to close the throttle of the number one engine.

(e) *Action.* This phase is the final part of the sequence and yet another source of error. The co-pilot might intend to move the throttle for the number one engine, but in fact moves that for the number two engine – a type of error too often compounded by poor ergonomic design. For if the controls are designed on a cosmetic rather than a functional basis, Murphy's law – which states that anything that can go wrong will go wrong – must be fulfilled.

In this case, however, the co-pilot closes the correct throttle and so complies with the order of his captain. This action sequence is concluded when the co-pilot's brain receives feedback information from his hand, and perhaps visual confirmation from the appropriate r.p.m. gauge, that the throttle has been closed. The complete sequence from command to action has probably taken less than a second; and yet it has contained at least five 'error-possible' points. The realisation that thousands of such sequences are completed on every trip brings one fact into a truly startling focus – namely, that gross errors are so few, and disastrous errors extremely rare.

2. *Visual illusions*

Most of the information which the pilot receives comes to him through his eyes. Some of this information comes from instrument displays in the cockpit, but a large amount is obtained from outside the cockpit, often under conditions which may be far from ideal. Indeed, certain conditions may prevent the necessary information even from reaching the eye – for example a pilot may fly into an obstruction obscured by cloud.

More often a signal reaches the eye, but the brain misinterprets it and the pilot 'sees' something else; in other words, he experiences a visual illusion. Such illusions are extremely common, and we all probably 'see' several of them each day – simply because the visual signal reacts with previous personal experience to produce the picture which is actually 'seen'. Thus experience teaches that objects usually maintain their size, so that when they appear to be getting larger – i.e. they form a larger image on the retina – they must in fact be getting closer; and behaviour is adjusted accordingly. A pilot makes thousands of such simple interpretations on every trip, but there are other more subtle illusions which require all his experience and concentration if he is not to be led astray by them.

The relationship between visual illusions and aircraft accidents has been excellently covered by Pitts (1967) who lists the causes of visual illusions commonly encountered in aviation. Among these are:

(a) *Refraction*. The curvature of the windscreen, or perhaps water on it, may cause refraction so that the pilot thinks that he is higher than he really is, and consequently lowers his approach.

(b) *Fog*. In fog, objects may appear to be larger and further away than usual, so tempting the pilot to fly below the glide path.

(c) *Texture*. The texture of the ground gives cues as to the observer's distance from it, while objects of unusual size, e.g. stunted trees or a particularly large building, may cause distance to be misjudged. Similarly, terrain which is nearly devoid of textural cues – such as still water – can make the judgement of distance extremely difficult. This is also true when there is little contrast between the runway and the surrounding terrain; 'degraded visual cues' because of snow was cited as one of the factors which caused the accident to a Viscount aircraft at Castle Donington in 1969 (see CAP 337).

(d) *Autokinetics*. A stationary light in a dark environment can sometimes appear to move. This can be demonstrated by staring fixedly at a faint point source of light in a dark room: after a while the light will appear to be moving around of its own volition. Similarly, when flying over uninhabited terrain on a dark night, a faint stationary light on the ground might be 'seen' to move and thus be misidentified as another aircraft.

One of the purposes of giving a captain familiarisation trips to an airport, before allowing him to approach it on his own, is to let him learn the vital cues which he will need for a visual approach. He will learn what sort of visual picture to expect at various stages in the approach and let-down, and what snares or illusions are waiting for him; e.g. that the forest on the top of the hill is composed of dwarf trees and so appears to be farther away than it really is; or that the dip in the ground before the start of the runway is far deeper than it looks from 2 miles out.

It is said that a pilot becomes 'tuned' to the visual picture provided by a three-degree glide slope, and provided that the visual cues on the approach to a particular airport are fairly typical, or are sufficiently plentiful to overcome any illusions, this tuning is both necessary and safe. When neither of these provisos is met, an accident situation may develop – as witness the circumstances highlighted by Kraft (1969). (See also Captain Bressey's description of the 'ideal image', p. 21.)

Kraft found that 16 per cent of all the major aircraft accidents to US carriers between 1959 and 1967 occurred during night approaches over unlighted terrain or water, towards well-lighted cities. He studied these accidents, setting up laboratory simulations of night approaches to certain airports, and found that a major cause of height misjudgement under these conditions was a descent path which nulled out visual information. The change in visual information from the airport lights during the approach was not sufficient to be perceived, nor were there other visual cues. Thus, the pilot had no visual indication from outside the cockpit that he was dropping below the glide path. On the basis that motion must exceed about one minute of visual angle per second to be perceived, Kraft calculated that at a speed of 240 miles an hour and at a height of 3000 ft – a typical approach configuration – such perception would occur at a distance of about 9 miles from touch-down; the distance at which several accidents have in fact occurred.

Such visual illusions are exacerbated by:

(a) Complete darkness both beneath and to the side of the approach.
(b) A long straight-in approach to an airport located on the near side of a city.
(c) A runway of unusual length or width.
(d) A sloping runway; i.e., the pilot assumes that the runway is horizontal and approaches it at an angle of 3°. If the runway slopes *up* at an angle of 3°, his approach may be disastrously low.
(e) A sprawling city with irregular lights on hills behind the runway; this provides the pilot with false cues as to height, distance, and the horizontal.

3. *The false hypothesis*

The false hypothesis is such an important contributor to human error that it deserves special attention.

For a simple demonstration of the characteristic thought processes involved, the reader should ask a friend to say the word 'boast' five times, as quickly as possible, and then to repeat the process with the name of the object he puts into the toaster. If he says 'toast' rather than 'bread' – as well he might, since his mind has become 'conditioned' to a phonic pattern – he has made a false hypothesis.

Such a vast amount of information impinges on the brain that it cannot possibly process it all in detail. The human reaction therefore is to take in only a small part of the information, and make an assumption about the rest. Typically, a pilot may see lights below him, assume that they are the lights of Airport A, and start to set up his landing pattern on this assumption. Usually they are indeed the lights of Airport A and his action is appropriate; but occasionally they are the lights of Airport B and he has made a false hypothesis which may lead to disaster.

On 14 January 1969 a BAC 1-11 crashed near Milan due to a loss of power, although both engines were subsequently found to be virtually fully serviceable.* Shortly after take-off a compressor bang surge occurred in the number two engine. For reasons which may well have

*CAP 347: *Report of the Italian Commission of Inquiry on the accident to BAC 1-11 G-ASJJ at S. Donato Milanese, Italy, 14 January 1969.* HMSO, London.

been rooted in fatigue, a training captain who was sitting in the jump-seat thereupon made the false hypothesis that it was the number one engine which had failed, and ordered that engine to be closed down. Unfortunately the thrust from the number one engine had already been partially reduced due to 'an inadvertent displacement of the relevant throttle lever', and the aircraft crashed. (See further reference to this accident, pp. 77–8.)

An early account of the 'false hypothesis' situation is provided by Davis (1958) who cites examples from the railways in which a driver looked at a red signal, read it as green, and caused an accident. Many pilots will doubtless recognise parallels in their own experience, and Davis describes four specific types of situation in which a false hypothesis is particularly likely to occur:

(a) *When expectancy is very high.* Long experience of an event which has always happened in a particular manner generates a strong probability that it will be 'seen', 'heard' or 'felt' to happen in that manner in the future, regardless of the precise nature of the stimulus. An example, in Britain, is the official blue and white sign which reads: 'Police Notice, No Parking'. A sign of similar design which contains the words 'Polite Notice, No Parking' is often used by harassed residents to encourage motorists to move on and park elsewhere. The stimulus received at the eye is the letter 't'; but because people are used to seeing a 'c' in this context, it is read as such.

Man has a great need to structure situations; that is to give them order and meaning. A pilot who hears a message over a distorted radio/telephone channel may make a number of hypotheses before asking for the message to be repeated. He knows that the airport tower would not pass a meaningless message to him, so he tries to fill in the gaps and believe that he heard what he thinks he *should* have heard. If he is expecting a particular message to be passed at that time, or if he is under great pressure – such as shortage of time or the intimidating influence of a domineering captain – this tendency to accept a dubious hypothesis is given even greater impetus.

(b) *When the hypothesis serves as a defence.* Human beings are prone to interpret incoming information in such a way as to minimise anxiety. It is, in fact, commonly witnessed that a person who is suffering from an incurable illness may readily believe an implausible hypothesis if it offers hope. And it is equally common for the student who receives a card informing him that he has failed an examination to assume that it has been sent to him in error. Both these people might be objective where the events pertain to others unknown to them; but they become irrational when they themselves, or people dear to them are concerned. The objective evidence is there, but people so affected have no wish to see it. Such disregard for reality is frequently evident when two people become involved in a violent argument. Both persist in advocating hypotheses which grow progressively less tenable because of the anxiety with which each disputant views the prospect of backing down.

However senior his rank, and however great his experience, a pilot is still subject to anxiety and, because of this, to the pitfalls of false hypothesis. Happily, his experience may have taught him appropriate lessons, and he may realise and guard against the circumstances which can lead to this type of error. Occasionally, however, his very seniority and status may be an even greater bar to admitting his mistakes – and he will only publicly reject his false hypothesis when it is too late, and when disaster is imminent.

(c) *When attention is elsewhere.* If a pilot has a number of immediate tasks, and if one of these requires special attention, he is likely to be less critical in accepting hypotheses about other components of the work load. Thus, if a great deal of the pilot's effort is devoted to spotting runway lights in bad weather, he is more likely to misread – that is, develop false hypotheses about – other instruments. Again, if the pilot *already* holds a false hypothesis about some item of information upon which he is concentrating very closely, he may ignore other vital information which conflicts with this hypothesis.

Davis (1958) cites as an example an accident in which the aircraft's starboard engine began to surge and lose power. The pilot reported to the tower that the propeller constant-speed unit had failed, and attempted to clear the surging by operating the throttle and pitch levers throughout their full range, but without effect. Shortly afterwards, the aircraft crashed.

The subsequent investigation showed that the cause of the accident was the single and simple fact that the fuel tanks were empty. But having made up his mind that the propeller constant-speed unit was defective, the pilot failed to carry out checks elsewhere in the cockpit.

(d) *After a period of high concentration.* Every pilot knows of 'end-of-tripitis' – that time at the end of a brilliantly flown sortie when the concentration drops. The most difficult part of the flying procedure has been successfully completed, but it is all too often at this juncture that an accident occurs.

Among the examples offered by Davis is that of a railway driver who drove through fog from Crewe to London – a distance of 160 miles. A few miles away from the terminus the fog cleared and the sun came out. The driver relaxed, drove through a red signal and crashed into another train. Clearly, the greater part of the trip – in poor visibility – had made great demands on his concentration. The remainder of the journey must have seemed so easy, in contrast, that the unfortunate driver had obviously considered himself 'as good as home'.

A false hypothesis need not be spontaneous but may be an idea with which a person has lived for many years. On a light aircraft such as the Chipmunk, the stalling speed of the clean aircraft is about 35 knots; this fact, too often, becomes a broad concept of the aeroplane's stalling characteristics. When the aircraft is in a turn with the flaps down, however, as may well be the case in an attempted forced landing into a field, the stalling speed is nearly doubled. In about one in two of all light aircraft accidents in the UK the penultimate act in the drama is a stall in the turn at low altitude. Virtually all light aircraft pilots know the stalling speed of their clean aircraft in straight and level flight, but how many know the other, and more vital figure? One who may not have done so was the pilot of a Nipper T66 who stalled in the turn at 200 ft. The figure for a power-on stall in this aircraft is 3£ knots, but becomes 65 knots at a 75° angle of bank.*

4. *Man, the creature of habit*

A basic tenet of human behaviour which is of vital importance to aviation is that the more times an action has been performed in a particular way in the past, the more likely it is to be performed in that way in the future. Indeed, much of the pilot's training is devoted to establishing a repertoire of habitual actions so that when he is flying he can concentrate on coordinating whole groups of actions, instead of devoting inordinate attention to one comparatively simple task. His basic role is therefore that of monitor of a number of low-level activities.

The majority of aircraft, too, are standardised, so that actions appropriate to one aircraft are also appropriate to another; i.e. controls are usually in the same place and operate – and are operated – in the same way. For the most part, such previously learned habits are beneficial,

* Civil Aviation Accident Report No. 8, 1973.

and may often be essential, to the present situation; but occasionally they are inappropriate and a potential accident looms large.

A much-quoted episode from the records of aviation medicine is provided by the case of the Auster aircraft which, shortly after take-off, was seen to porpoise and dive into the ground (Allnutt, 1971). The pilot had applied excessive elevator trim, probably because he had been flying a different aeroplane – a Tripacer – for much of the preceding year. Both the Auster and the Tripacer are fitted with a similar trim control in the same place, but whereas the Tripacer requires several turns of the control wheel to effect a small change in trim, the Auster needs only a single turn to cover the full range. The pilot probably carried out the right action for a Tripacer; unfortunately, he was in an Auster at the time.

This accident, and others like it, illuminate one great disadvantage of the human system when it is compared with a machine: namely, that man has no erase button on his memory. *He cannot make himself forget.* Rather, old and no longer appropriate habits must be slowly and laboriously eliminated, whenever man must adapt to new conditions and demands. This situation can be exacerbated in certain circumstances when people are more likely to revert to earlier and well-learned patterns of behaviour.

It is known that two situations of this type encourage such reversion – firstly, when little attention is being paid to the task in hand. An example is the familiar experience of the driver who changes from a floor-mounted gear lever in a car, to one which is mounted on the steering column. While the driver concentrates on changing gear, his actions are appropriate; but as soon as he attempts to change gear with his attention elsewhere he finds his hand waving about at knee level.

A second cause of reversion to earlier behaviour patterns is stress. It can be shown in the laboratory that a person under stress is much more likely to repeat an earlier, and possibly inappropriate action.

The possibility of such an occurrence is suggested in the Report of the Italian Commission of Inquiry which investigated the crash of the BAC 1-11 aircraft referred to earlier. The report states

> Psychologists consulted think that there may have been a return to a former behaviour pattern acquired during operation of aircraft types which required immediate action by the pilot in the case of an engine failure on take-off.
>
> During the nine years preceding his service in the BAC 1-11 (the pilot)

did in fact fly piston-engined and propeller turbine multi-engined aircraft and during that period he experienced an engine failure on take-off.

Investigators concerned with specific cases of pilot error must therefore consider not only the pilot's present habits, but also those of his past. It is not always possible to assess the relevance of these patterns, or even to pronounce with any accuracy that an old and inappropriate habit has ceased to be a danger. This depends both on the strength of the original habit and on the subject's subsequent behaviour patterns. Suffice it to say that all habits go very deep, and lack of attention – or the prevalence of stress – may awaken a habit which has lain dormant for months, *or even years.*

As a pilot's experience increases, so does his repertoire of responses and so, alas, does his age. Questions on the relationship between accident rates and age and experience are often posed, and are rarely answered satisfactorily: but it requires little effort to appreciate how complex this relationship is. Age and experience and their corollaries are usually correlated; i.e. the older pilot is frequently more experienced than his younger colleague, but may be in a slightly worse physical condition, although he may counter the latter effect by pacing himself more efficiently. He may be more worried about status, but will have learned to live with many of life's other problems. The list of such compensatory adjustments is endless. Yet in a simplistic review of age and experience it is true to say that in the context of physical fitness the human body reaches a peak in the years of the 'teens and early twenties, and from then on the progress is downhill; whereas relevant (and irrelevant) experience increases throughout life.

The ideal, and probably the safest, age for any job depends upon the nature of that job. A female swimmer usually reaches her peak in her late 'teens, whereas a High Court judge is usually in his late fifties or may be even older. Similarly the fighter pilot who has a hard physical task which needs good reactions may be at his best in his late twenties, while the captain of a large aircraft who has a large management element in his job may reach his peak somewhat later. Graphs purporting to show the relationship between accident rates and age and experience should always be considered with these provisos in mind.

Recent experience, i.e. current familiarity with the aircraft type, is an important factor in flight safety; for statistics show that the pilot who flies less than 100 hours per year is the more likely to have an accident (Smith 1966). Current acquaintance with specific manœuvres is of equally crucial importance – a fact confirmed by such events as the crash of a twin-engined Apache which came to grief after one of the engines failed. The investigators could find no record that the pilot had practised asymmetric flying at any time during the previous twelve years.*

A large proportion of the professional pilot's training, however, is devoted to preparing him for situations which it is hoped he will never meet; that is, for real emergencies. The object of this training is two-fold. Firstly, it is to ensure that the pilot's responses to the emergency are made as automatic as possible in order to 'free' his mind to a great extent, for concentration on the crisis; in other words, the pilot 'over-learns' the emergency-response sequence. This process of over-learning is essential, since the shock of an emergency may well diminish the pilot's ability to a considerable degree. Secondly, over-learned responses and realistic simulations give the pilot the confidence to remain calm. In actual emergencies, however, responses are likely to be elicited by less intense and less specific stimuli, and tend to be more forceful, more extensive, and more rapid; at the same time they are somewhat less well organised, less regular, and in all probability, uncoordinated.

However realistic the simulator, the emotional shock of a genuine emergency is missing; and so the margin of over-learning of the safety procedure must be large. The ability to climb slowly into a rubber dinghy in a warm swimming pool is of little use; what counts is to be able to do it quickly in darkness, and in freezing water, while nursing a bruised head and rehearsing an exoneratory story for the Board of Enquiry. In the event, no responsible pilot neglects his emergency training and all of them would endorse the quotation from the UK's National Safety Council: 'No-one gets ready for an emergency in a moment. What a person does in an emergency is determined by what he has been doing regularly for a long time.'

5. *Motivation*

One glib definition of the difference between a man and a machine is that a man cares about getting home safely, whereas a machine does not. Apart from the occasional aviation suicide, this is true; and so motivation must be seen as a major factor in any consideration

* CAP 342. Report on Accident to Piper PA-23 (Apache) G-ARHJ at Hilfield Reservoir, near Elstree Aerodrome, Herts, 27 January 1968.

of pilot error. The investigator must therefore consider not only whether the pilot had the ability to carry out a particular task, but also whether he was sufficiently motivated to do so.

This is not, of course, the task of getting home safely, but rather the thousands of component tasks which are carried out on every trip. Any failure to carry out even one of these tasks successfully may lead to an accident.

Exploration of the many elements of motivation represents a profound field of research. It may, however, be convenient here to consider motivation as one of the major components of arousal. There is, unfortunately, no single agreed definition or measure of arousal, but basically it is the general state of human alertness as shown by certain physiological and psychological indicators. At any one time, man's level of arousal varies on a continuum from deep sleep, up through the various states of wakefulness, to a state of blind panic. His level of arousal is determined by many internal and external forces, and of these perhaps one of the most compelling internal forces is his motivation. Two major aspects of motivation are level and direction:

(a) *Level of motivation.* It is a simple, but quite useful, dictum that performance is best at moderate levels of arousal. If arousal is too low, energy is insufficient and nothing is achieved; if arousal is too high, there may be plenty of action but it is ill-directed and useless. This 'day-dream-to-panic' continuum is described by Lager (1973) as a situation in which the brain is 'switched out' at low and high levels of arousal. At a low level the effect is analogous to low-level programs (in the computer sense) becoming muddled up, while at the high end a man may display random, i.e. purposeless, activity. Alternatively, at the high-arousal end, the brain's channel-selector may become jammed, causing the person to fixate on one thing to the exclusion of all others. As a typical consequence of this condition the passenger trapped in the blazing wreckage of an aircraft may continue to struggle with an unyielding emergency door and ignore the gaping hole in the fuselage. Given the emergency situation without the fire, and consequently a much lower level of arousal, he would easily spot the alternative exit.

Although it is the panic end of the arousal continuum which is the more spectacular, it is often the other end which is the more dangerous, for at the low end of the continuum lies complacency. Aircraft have crashed into each other in clear skies, experienced pilots have landed wheels-up, and many pilots have begun what proved to be a final flight secure in the knowledge that 'it can't happen to me'.

Complacency is ubiquitous and not easy to prevent. Artificial signals to raise the pilot's arousal to an optimum level are of limited use because he soon becomes adapted to the signal. It has been found, for example, that it is not possible to prevent wheels-up landings by burdening the pilots with the onus of an additional call to say that the under-carriage is down; for this call soon becomes part of the routine, and thus non-arousing. The only real answer lies in education and pro-paganda so that every pilot is always aware that every safe landing he makes is, on the complacency index, making matters worse.

There have been numerous accidents which can be described as having been due either to too high or too low a level of arousal, but for many years to come researchers will have to remain content with descriptions rather than proof. At the moment it is only possible to hypothesise, and to use evidence from witnesses or voice-recordings to estimate a pilot's level of arousal at the time of an accident. Lager (1973) has tried to close the gap between description and proof by measuring pilot performance in a flight simulator. He reports that he can distinguish groups of pilots who possess either a very low or a very high level of arousal, as measured by certain physiological indicants. Both groups perform significantly worse than the pilot population as a whole, and also have a poorer – i.e. more frequent – accident record. The low-arousal group show very little emotion after an accident or a poor trip in the simulator; while the high-level group are very active, but subject to performance breakdown if energy alone cannot solve their problem.

The levels of motivation which people appear to experience vary greatly. There are numerous stories of people who have exerted a superhuman effort in moments of great stress: unaided they have lifted a huge weight from a trapped friend, or suffered dreadful hard-ships at the behest of that supreme motivator, survival. But be it noted that virtually all these actions are fairly simple physical acts; what is more rare is the demonstration of a balanced and rational approach to a problem when the motivation to succeed is very high – a situation exemplified by the trauma of examination nerves.

(b) *Direction of motivation.* To say that a pilot is motivated merely to fly from place *A* to place *B* is to over-simplify the associated problems. The pilot's decision is not whether to fly safely or to have an accident; but often whether to take very slight risks, or just slight risks. In this context there are inevitably conflicting pressures on a pilot who is about to make a landing in marginal weather conditions. The decision is his and he knows he *could*, and perhaps *should*, divert to an airport some distance away. On the other hand, his mind is processing the counter-arguments: 'The company would be glad if I managed to land there; the passengers are waiting; I'd like to get home tonight; and I don't want to miss my golf/theatre/dinner appointment ...'*

One factor in particular may play a major part in such a dialogue; yet in any discussion the issue of manhood versus safety is usually swept under the carpet with alacrity.

Our culture and training teach us that men are brave; that it is natural, desirable and indeed heroic to struggle through adversities such as pain, bullies, strong teams of opposing sportsmen, and enemy troops. Man's lowly instincts therefore often generate a contempt for bad weather, or a risky flight, as just another hazard which will yield to bravery. This distortion of logic is compounded by social attitudes offering a combination of censure and sneaking admiration for the man who breaks the rules and wins; while of course, the wrath of righteous indignation is called down upon the man who breaks the rules and loses. But most, if not all, men have the occasional desire to kick over the traces and to take risks they would not ordinarily take, as the accident statistics on young drivers confirm. For them, as for many other people, the safe course often carries with it a slight, but nevertheless perceptible loss of face; and 'face' – as the antithesis of weakness – must be preserved.

Pilots are trained to control the urge to express 'manhood' by flying dangerously, and the urge *is* controlled, but not entirely destroyed. Mason (1972) quotes the case of a Convair Metropolitan aircraft which came in to land at a small, fog-shrouded coastal town. The weather was below limits but the pilot succeeded in landing at his third attempt. A second Convair approached and was told by the airport Tower that the first aircraft was already on the ground – information which clearly represented sufficient challenge to the incoming pilot to sustain him through two abortive approaches before his aircraft crashed into a group of houses, killing 41 people.†

* Cf. the description of decision-making in marginal weather conditions which Captain Price offers in Chapter 6, pp. 210–12.
† Cf. findings of CAP 270, noted in Chapter 6, p. 213.

6. *Stress and stressful environments*

Man is equipped with a number of defence mechanisms which counter the effects of any stressful environment in which he may find himself. If he is too hot therefore, he sweats; if he is tired, he may try harder; and if he is feeling the effects of age, he may adopt a more philosophical approach to life. But for all such stresses there may come a time when the defence mechanism can no longer cope adequately with the stress, and performance suffers. The degree to which pilots can be so affected, and in what manner, is a matter calling for some clarification of the stresses to which he is exposed.

For convenience, three categories of stress may be identified: firstly, the physical stresses such as temperature, vibration, turbulence, and lack of humidity; secondly, the physiological stresses such as sleep-loss or disturbance, irregularity in eating patterns and the effects of drugs such as alcohol, nicotine, etc.; and thirdly, the psychological stresses such as fear, frustration and social and commercial pressures. By and large, research workers know more about the effects of physical stresses than they do about those which are physiological; and more about those which are physiological than about those rooted in psychology.

In the aftermath of an accident it may be possible to establish the presence and approximate magnitude of the physical and physiological stresses; but it is far harder to do the same for the psychological element. In a crash investigation, for example, the acceleration to which the cockpit was subjected might have been recorded and enquiries can reveal the pilot's sleep patterns for the previous few days. But establishing the importance of the real relationship between the crew members as a factor in the accident, or the degree to which the captain was worrying about his child's health or his own possible impotency, may well be virtually impossible.

The situation is the same in the laboratory, for stresses such as are imposed by large variations in temperature or by an excess of alcohol are measurable and reasonably well documented. It is possible to say that if, for instance, the body temperature exceeds $38 \cdot 0°C$, then human performance may change in the direction of increased speed and tendency to error, together with diminished accuracy and greater irregularity. Equally, more than 80 milligrams of alcohol in a blood sample convinces the police that a car driver's erratic performance was influenced by drink. Unfortunately certain stresses are not so well documented as others, and for many of those which are of the greatest interest to commercial aviation – such as disturbances in eating and sleeping patterns, low humidity, conflict and frustration – the data on performance effects are very limited. Yet, even when the better-documented stresses are relevant, there is still a wide gulf between the laboratory experiment and the responsibility of providing a definitive answer to the question posed at an accident enquiry: 'Did the presence of this stress contribute to the accident?' The question, in turn poses others for the competent investigator, who would want to know a good deal more about the nature of the stress, the work which the pilot had to do – and about the pilot himself.

Besides the severity of the stress, the investigator would want to know its duration; the suddenness of its onset (a factor of considerable importance) and the presence of other stresses. Thus, one night of poor sleep may have effects very different from a week of such nights; and gradual adaptation to a hot climate is less deleterious than a sudden change of environment. Many (not quite all) of the data on the effects of stress have been obtained using single stresses whereas real-life situations almost invariably contain a number of stresses. To predict the effects of a multi-stress environment from the known effects of its component stresses is a most unreliable procedure, however, for the stresses may combine in an additive, more-than-additive, independent or even antagonistic manner. The effects of some of the vast number of stresses which plague modern man have been reviewed by many authors, among the more recent being Poulton (1971) and Appley and Trumbull (1967).

Both the nature and difficulty of the task are important variables, as the pilot may well be able to cope with the associated stress when the flight is routine, but cannot manage when events begin to pile up on him. A 'chapter of accidents' is a common description of events in which a number of minor difficulties has occurred – either simultaneously and/or progressively – and their combined effect has made the pilot's task impossible. A further complication is posed by the fact that man is often a very bad judge of how well or badly he is performing a task – as witness the drunken driver who drives badly, while believing that he is driving well. Similarly, a senior pilot who is angry because his flying is suffering from the effect of stress may not be over-receptive to remarks to this effect from his juniors.

Lastly, the effects of stress depend on the pilot himself, his physical fitness, his competence and his personality; and here again the relationship between these factors challenges positive affirmation. Every reader knows people who have coped with stresses in far better fashion than would ever have been predicted of them. Men such as Byrd, Bombard, Lovell and hundreds of nameless pilots have responded magnificently to great stresses; others have failed at the first hurdle. The scientific selection of stress-resistant people is still in its infancy, although it has proved successful in the astronaut selection programme and is now in use with an increasing number of organisations. But, in general, it is true to say that the study of the physical and physiological stresses which affect a pilot is still in a state of virtual adolescence, and the picture of what happens when man is subjected to these pressures is being completed only slowly.

In recent years, psychological or 'life' stresses have been receiving a great deal more attention, as their ubiquity and great importance are being recognised. A pilot may say that he does not allow his work and his domestic life to mix; but this statement can only be partly true. Human beings are 24-hour-a-day people, possessing only one brain with which to control all their activities; and this brain has to cover both work and play. In sum, events which happen in one segment of daily life may therefore influence what happens in other segments.

The pilot who has just quarrelled violently is in a dangerous state,* for although he may have moved away from the person with whom he has quarrelled, and climbed aboard his aircraft, the physiological and psychological effects of the quarrel may last well into the flight, and the crushing retort which he wishes he had thought of at the time of the argument may crowd his single decision channel to the exclusion of more important information.

Few discussions of pilot error proceed very far before mention is made of 'accident-proneness' – the idea that some pilots are inherently more likely to have accidents than are others. This concept comes in and out of fashion as the years pass and as new studies are published (see Appendix, p. 257), but at best, the thesis is of very limited use in an aviation context for the following reasons:

(a) The number of aviation accidents is small and so data are scarce.

(b) If accidents are normally distributed amongst the pilot population there are, quite naturally, pilots who have had more than one accident.

(c) Commercial and military pilots are very highly selected and what

* See reference to the Trident Enquiry in Chapter 1, pp. 60–1.

may apply, say, to the motoring public at large, probably does not apply to such a small select group.

(d) Even were the concept of accident-proneness to be proven, there is little practical action which can be taken. By the time that it had ꞵeen shown that the chances of a particular pilot having an accident ꞁad increased from one per million units to five per million units a great deal of money will have been spent on his training.* Singling him out for special treatment – restriction to good weather flying or to daylight flying only – would, of course, be disastrous for his own and his colleagues' morale.

A much more useful and perhaps more realistic concept than accident proneness is that of a short-term accident liability – the proposition that a pilot becomes more likely to experience an accident because of the personal pressures on him at the time. More specifically, in this attempt to rationalise cause and effect, the man whose child is ill, or who has quarrelled with his wife, or who is worried about his job, may be more likely to have an accident until he emerges from the trauma which is disturbing him. It follows, of course, that given an appreciation of this situation, the responsible pilot or manager should appreciate these stresses and make the necessary adjustments.

7. *Fatigue*

Fatigue might have been considered with the other stresses to which a pilot is subjected; but it is a topic which is discussed so often in the pilot-error context that it deserves special attention.

Captain Bressey is correct in his claim that '... there is still no definition of fatigue acceptable to all interested parties; nor has medical science been able to define, in specific terms, what produces fatigue ... (Chapter 1, p. 27).

Nevertheless, definitions have been attempted and many of them may be subsumed under that of Bergin (1948):

> Fatigue is a progressive decline in man's ability to carry out his appointed task which may become apparent through deterioration in the quality of work, lack of enthusiasm, inaccuracy, ennui, disinterestedness, a falling back in achievement or some other more indefinable symptoms.

A distinction is sometimes made between single-trip fatigue, where a good night's sleep will effect recovery, cumulative fatigue, in which a

---

*See the parallels in Chapter 3, pp. 93–4. 121, 130–1.

state of rest between trips is not adequate and which may necessitate several days' rest for complete recovery, and chronic fatigue, which usually calls for medical help. The behavioural symptoms of all three types of fatigue are similar, however, and usually include increased irritability and irrational behaviour, as well as complaints about physical discomfort, the whole accompanied by a deterioration in flying skill. Thus, at the end of a tiring day's flying a pilot may perform less accurately and attentively, may require a higher than normal level of stimulus to evoke the appropriate responses, and may show increased fixation of both perception and behaviour.

A plethora of facts hasten the onset of fatigue. Organisational factors, for instance, may include badly planned or delayed trips, unnecessary delays, unproductive time or inadequate stop-over facilities. Any or all of these factors may be compounded by poor ergonomics in the cockpit, high work-load, temperature, noise, vibration, low humidity and bad weather. These pressures may be intensified by the physiological factors, by any decline in the pilot's fitness, or by the effects of mild hypoxia (oxygen deficiency), lack of food or disturbed diurnal rhythms. They may be even further intensified by psychological factors such as responsibility, morale, crew cooperation, and worry.

All of these, and many other factors relevant to aviation fatigue, are described in an excellent work by Hartmann (1969).

Most people connected with flying recognise the symptoms of fatigue both in themselves and in others. It remains, nevertheless, difficult to provide a quantitative measure of the condition and nearly impossible to prove that it may have caused an accident. Post-flight psychological and hormonal measures of fatigue have met with only limited success, since fatigue is usually only one of a number of influences on these measures. Others, such as the attitude of a tired pilot in having to carry out a 'damn-fool' psychological task at the end of a long day, may be equally important.

Post-Accident Information

A variety of sources may provide post-accident clues to human behaviour. Tape recordings of communications between the pilots, or between the pilots and the ground, may tell the investigators just what was said, and the manner in which it was said – the latter facet perhaps indicating the presence of stress. Unfortunately the tapes may give no information about the pilot's plans or fears, nor may they give any indication as to whether or not he had actually seen a particular warning light. Experts, too, may not agree that the pitch of a pilot's voice is necessarily a useful index of stress.

A second major source of information is provided by the evidence of the pilot, and of other parties involved in the accident; and here there are many possible sources of distortion. Occasionally one or more people may lie, either in an attempt to save themselves, to avoid embarrassment, to protect their jobs, or out of misplaced loyalty to a colleague. The word 'misplaced' is used with some reluctance, but it is clear that the vital necessity for total honesty and objectivity in accident investigation may well conflict with some very deep emotions.

A far more important source of distortion lies in the poorness of the human memory. Man absorbs a truly vast amount of information during every moment of his life. Much of this remains in the memory for a very short time; a very little dwells for longer periods. But every moment, memories are decaying and are being contaminated with the arrival of fresh information. A pilot may, therefore, remember the pressure setting he used on the trip from which he had just landed, but probably does not remember what value he set last Tuesday week; and the thousands of such settings which he has made over the years form a general blur.

Closely controlled laboratory experiments confirm the everyday experience that memories are distorted in the direction of simplicity and coherence. If a story is repeated many times from one witness to another, it is progressively simplified until only the bare bones remain. Similarly, the human desire for order out of chaos means that non-sequiturs in a story are gradually eliminated until a clear and logical account of events emerges. However, this final account usually bears very little resemblance to the original message.

The pilot who repeats his story a number of times, and who discusses the accident with many people, may, with total integrity, give a false account of what actually happened. He 'remembers' that the reading on a dial was such and such because that reading would make sense in the context of his story.

Distortion can be minimised, although not eliminated, when statements are taken from the parties concerned as soon as possible after the accident. However, pilots have yet to be trained to crash alongside the investigator's office, and prevention of any discussion of the accident pending the arrival of the investigator is not without its difficulties.

It is an unfortunate fact, too, that the investigator himself may be a source of distortion, for the answers to his questions depend, at least in part, on the questions he asks and the way in which he asks them. Every investigator has his personal bias, and a pilot, a lawyer and a psychologist in search of a common truth may ask very different questions and receive very different answers.

The non-directed interview, in which the witness talks and the interviewer merely records his statement, goes some way towards solving this problem. Eventually, when the enquiry is completed, the pilot-error cause is coded and entries appear against certain categories in a record card. This information is handed down to posterity, but again, may be a distorted version of what actually occurred.

Finally, eye-witness accounts may provide a little cheer in the tragic circumstances of an accident, since such witnesses are often victims of many of the pressures mentioned above. They may have talked to a number of people about the accident; they have almost certainly read graphic accounts of it in the newspapers, and they are usually motivated by a strong desire to appear both knowledgeable and helpful. 'The port engine backfired and the aircraft yawed sharply to port in a marked nose-up attitude – I make model aircraft so I know about these things.' This may appear to be a helpful statement but the investigator has to satisfy himself that the witness really did understand the meaning of each term, and, what is more important, that he understood them at the time of the accident. The tendency for distortion in the direction of what *must* have happened rather than what *did* happen must not be forgotten. If a piston-engined aircraft passed overhead, appearing to be both 'low and slow' before it subsequently crashed, an obvious corollary for the lay observer is that an engine must have been running roughly.*

It is salutary to conclude this chapter by reiterating that mistakes are a normal feature of human behaviour and that aviation is a human activity. It is nevertheless equally human that the logic of this

---

* This syndrome was plainly in evidence following the Moorgate tube-train crash in London on 28 February 1975, in which 41 passengers were killed.

Following conjecture in the press that the driver had suffered a heart attack (refuted by subsequent medical evidence) one British newspaper reported as follows:

Eye-witness reports – taken by British Transport police – of (the driver's) last moments at the controls of the train indicate that he was a sick man. One said: 'The driver looked as if he were mesmerised. He was staring in front of him and just looking ahead.'
Another said that the driver had a blank expression.
  A statement from a passenger waiting to board the train said ... 'the driver had a glazed look in his eyes and looked as though he was frozen'.   (*Daily Telegraph*, 5 March 1975)

proposition should be challenged by the continuing pursuit of twin goals – firstly, the reduction of the possibilities for pilot error; and secondly, the lessening of the consequences of such mistakes.*

The achievement of these goals, however, depends on much more than an acceleration of the present trend towards a better understanding of pilot behaviour. It requires, in addition, a far more rational attitude towards their mistakes. That 'better understanding' is essential for all those concerned with aviation safety: for pilots, managers, the designers of aircraft and procedures, those responsible for the selection and training of pilots, research workers, accident investigators and many others. For all of these, the data-producers must present their evidence in a usable format wherein research findings and experience are collated and state-of-the-art reviews are readily accessible. The decision-makers must seek these data and use them. But, above all, the attitude to pilot error must be unemotional and constructive; for until the pilot's errors are viewed as dispassionately as are those of the aircraft in which he is flying, there will be more than a ring of truth in the cynical phrase: 'If the accident doesn't kill the pilot, the enquiry will.'

References

Allnutt, M. F. (1971) 'Human factors: their significance in an investigation'. Paper presented to the D.T.I. seminar on Aircraft Accidents, London.

Appley, M. H. and Trumbull, R. (1967) *Psychological Stress*, Appleton-Century-Crofts, New York.

Bergin, K. G. (1948) *Textbook of Aviation Medicine*, J. Wright, London.

David, R. D. (1958) 'Human engineering in transportation accidents', *Ergonomics*, **2**, 24–33.

Hartman, B. O. (1969) 'Psychological factors in flying fatigue', *International Psychiatric Clinics*. Winter (4)1, 185–96.

Kraft, C. L. (1969) 'Measurement of height and distance information provided to the pilot by the extra-cockpit visual', *M.I.T. Conf. Proc.*, 257–64, April 16–19.

Lager, C. G. (1973) 'Human factors.' Paper presented to a seminar on

---

*For examples of the type of area in which such 'reductions' might be made, see Chapter 4, pp. 153–60.
† See relevant comments in Chapter 3, pp. 107, 126. Comment in Chapter 6, pp. 247–8.

Aviation Accident Prevention at the Royal Institute of Technology, Stockholm.

McFarland, R. A. (1953) *Human Factors in Air Transportation*, McGraw-Hill, New York.

Mason, C. D. (1972) 'Manhood versus safety'. Paper presented to the Third Oriental Airlines Association Flight Safety Seminar, Singapore.

Masters, R. L. (1972) 'Analysis of pilot error-related aircraft accidents'. Paper submitted to the National Transportation Safety Board.

Pitts, D. G. (1967) *Visual Illusions and Aircraft Accidents.* Technical Report No. 28, School of Aviation Medicine, Brooks Air Force Base, Texas.

Poulton, E. C. (1971) *Environment and Human Efficiency*, Charles C. Thomas, Illinois.

Rolfe, J. M. (1972) 'Ergonomics and air safety'. *Applied Ergonomics* **3**(2), 75–81.

Shaw, L. and Sichel, H. (1971) *Accident Proneness*, Pergamon, Oxford.

Smith, E. M. B. (1966) 'Pilot error and aircraft accidents', *ZBL Verkehon-Med.* **12**, 1–13.

Wansbeek, G. C. (1969) 'Human factors in airline incidents'. Paper presented to the 22nd Annual International Air Safety Seminar at Montreux.

3

Designer's View
John Allen

The very considerable publicity given to real or alleged cases of pilot error in recent years has firmly entrenched the phenomenon in the lay mind as an inherent characteristic of air transport.

Yet parallels which illustrate the narrowness of this view can be drawn from pre-aviation history; and nowhere, indeed, are they more strikingly apparent than in K. C. Barnaby's remarkable book *Some Ship Disasters and Their Causes* (1968). This work recorded and interpreted nearly 100 cases of ship collision and grounding, largely taken from the nineteenth century. The singular circumstances leading up to the accidents described were clearly recalled from the official enquiry records, and it is from these that a familiar pattern emerges; namely, the frequency with which the ship's captain was accorded a manifestly unjust proportion of the blame.

The pattern is directly analogous with the situation of the airline pilot who is considered in this book, for although the legal instruments and financial pressures of the previous age are now eclipsed – in the first case in their range of opportunity, and in the second, in sheer scale and ramification – there is a consistency in the philosophy which shaped the fiercely autocratic judgements of the Admirals who conducted those Courts of Enquiry. In their opinion, too, command – i.e. the status of Captain – is equated with absolute responsibility, *regardless of inadequate design or provision.*

With wider experience the critical nature of the latter factors has been, and continues to be, established. The statement is carefully qualified, since the interface of man, machine and environment is by no means yet fully developed, and it is with this knowledge that the aircraft designer must approach his task. Clearly, any human controller of any public transport vehicle can and may contribute errors from time to time which, in turn, lead to fatalities and searching enquiries. The trauma for aviation, however, is that the events are usually more dramatic, and cannot but attract sensational press reaction.

The accident records of the major transport systems may be compared as follows:

Type of accident	Total deaths
Motor vehicle	52 924
Aircraft	1 799
Railway	997

These values apply to the USA in 1967. The exposure to risk in using a particular form of transport is proportional to the total product of passengers multiplied by the distance travelled. The accident rate is then the total number of deaths divided by the total passenger miles. The airlines figure is about 0·3 deaths per hundred million passenger miles; that for the motor vehicle, about 4 deaths per hundred million passenger miles.

For comparison, the deaths from other causes of accident are, over the corresponding period: falls, 20 120; fire and explosions, 7423; firearms, 2896; electric current, 376; lightning, 88; explosion of pressure vessels, 42.

No transport system can afford to allow accident rates to increase beyond norms such as those tabulated above, and, bearing in mind the continuing trend of improvement in aviation, its record is good.

In this chapter it is the scientific and technical aspects which are emphasised; but it is equally necessary to emphasise that complete understanding of the subject, and of the part played by pilot error, is dogged by four major issues:

(a) those unknown and therefore unmeasurable minutiae contributing to any actual accident;
(b) the practical impossibility of incorporating, after an airliner is in service, the very many possible changes which may improve safety;
(c) the impossibility of obtaining statistically certain evidence that

any particular 'improvement' will actually lead to worthwhile gains of safety;

(d) the fact that steps taken as safeguards against one kind of accident may help to promote one of another kind.

Consider first the degree of knowledge and certainty of various technical features of aviation as determined scientifically in actual special tests, and the very different state of ignorance and gross uncertainty which is bound to exist at the time of many an accident.

The contrast is perhaps not so surprising, for common sense will acknowledge that a similar problem exists in relation to the motor car. In an uncomplicated and straightforward situation such as a road test, power, acceleration, speed, turning ability and braking power can all be measured and identified. Yet how many essential facts and measurements are totally unknown in ascribing blame for high speed accidents which occur during relatively simple manœuvres such as overtaking? It must be clear, therefore, that the unknowns for the aeroplane must be many orders of magnitude more complex because of the far larger number of controls required for flying operations, the greater range of speed and acceleration, and the fact of movement in a third dimension.

It is this vast area of inevitable uncertainty and lack of pertinent data which leaves the residual doubt of the exact cause of an accident all too commonly ascribed to pilot error. The explanations can become distorted, patently incorrect, and often as patently unjust, in that influences such as profitability, competition, questions of legal compensation and national or professional prestige have entered the arena. Such influences, 'non-technical' though they may be, nevertheless make play within the margins of uncertainty and thus compound a fearsome battle for resolution from which the pilot must so often stand aside in dumbfounded bewilderment.* Yet a body of knowledge exists and it is drawn on here to examine those elements of air accidents which can be described and measured in scientific terms. Without such quantification there is certainly no hope for the achievement of higher standards of safety, nor for the success of the methods for its improvement adopted by all concerned with making and using aircraft.

* Cf. Comments in Chapter 1 on public enquiries (p. 60) and on the 'German authorities' (p. 53).

How air accidents happen

There are three basic causes of an air crash: firstly, a collision with a fixed object on the ground while the aircraft is in fully controlled and powered flight; secondly, a mid-air collision with another aircraft; and thirdly, a loss of control, lift or power, which in flight leads to a dive to the ground, and which on take-off – in any case of subsequent failure to stop, or to become airborne in the runway length available – leads to collision with off-runway objects.

The first type of accident may arise from an error of navigation, i.e. the pilot believed he was at a place or height different from his actual position, or that a fixed ground object, i.e. a radio mast, or mountain peak, was at a different relative position to himself. These errors may stem from malfunctioning instruments, or from the incorrect reading of instruments, or from incorrect deductions by the navigator.

The mid-air collision is broadly of similar origin, but here the errors of position of two aircraft combine to reduce safety separation distances to zero. Thus not only may the pilot's information be faulty but errors may arise in the air traffic control system, i.e. in the information given – or not given – by ground-based radars and computers or by the air traffic supervisors.

The third category of crash is quite different, as it depends only on the behaviour of the aircraft itself under any of the circumstances of loss of propulsion, loss of lift or control, or structural failure. A loss of propulsion may arise from mechanical defect, mishandling of the engine or fuel controls, or from a shortage of fuel. The former contingency cannot be ascribed to pilot error; the latter two may. Loss of control may result from mechanical or electrical faults, or from the pilot allowing the aircraft to reach excessive angles relative to the airflow, and so destroying the aerodynamic forces giving lift and stability.

The final factor in this category – structural failure – may arise from undetected corrosion weakening a vital part or as the result of a fire which has the same effect, or from severe mishandling of the aircraft during recovery from violent manœuvres – e.g., as in the attempt to recover from the effect of sudden avoiding action. Structural failure in these circumstances is a remote occurrence, but might correctly be ascribed to pilot error – at least in the sense that it was the pilot himself who activated the catalytic action.

Designing for safety

To see how safety is built into an aircraft to minimise the possibility of such accidents, it is necessary to probe in some detail into the origin and evolution of a new design; typically, that for a civil transport.

The major stages in the creation of an aircraft are related together in a framework within which a great variety of choices must be taken, and an equally significant level of compromise arrived at. It is in establishing these decisions – albeit that they are taken in the light of the best knowledge available to the manufacturer – *that the seeds may be sown for future errors and accidents.*

During the design process of the new type a vast number of alternative combinations and sizes of engine, cabin layout and equipment are studied – a process in which the airline operator, the manufacturer, national safety authorities and outside authorities (possibly including a Government with a financial interest in the new aircraft) are intimately involved. Compromises are struck between the advantages of significant new solutions such as improved engines or more automatic control systems, and the risk, cost and time of developing such new devices. As the time for signing the contract approaches the design becomes 'frozen' as all the basic principles and much of the detail is settled, hopefully for the last time. The means of achieving safety are often locked in at this stage by such matters as the inherent reliability of the actual mechanical and electrical systems chosen, and the particular stalling characteristics of the wing.* Subsequently, as the design proceeds into further detail (a process frequently requiring several million man hours and the completion of several tens of thousands of drawings of new parts) certain further decisions may be resolved which were indefinable at the contract-signing stage.

Some years later the first prototype emerges for its first flight. By this time the structure, the systems, and the engine will have been subjected to many hundreds and even thousands of ground tests which approach by various degrees of approximation conditions actually experienced in flight. A great deal still has to be proved, however, which can only be done by flight tests; and so the end of the prototype construction is the beginning of a period of flight test and certification which may take thousands of flight hours on a fleet of up to six aircraft, and last from one to five years.

* See case of Captain Foote, Chapter 1, pp. 46–7.

Because aircraft of a given type will fly very many millions of hours in the course of a life spanning many decades, the test flight period accumulating only a few thousand hours seems intolerably brief. Nevertheless. such flight programmes run up bills of scores of millions of dollars. Considerable judgement is needed, therefore, to decide on the scope of the testing, i.e. not only which performance and handling characteristics are to be measured, but which combinations of adverse factors should also be included – such as wind gusts, rainy weather, and take-offs and landings in cross winds and on icy runways. A range of measurements is made, too, of the aircraft's behaviour when certain systems are intentionally caused to malfunction within prescribed limits.

The type is finally approved by the safety authorities, and a certification document issued, stipulating all the associated conditions of loading, centre of gravity position, etc., for which the approval is given. This document is of primary importance, containing as it does measured information and defining limits beyond which the aircraft should not be operated. It is true that the question has often been asked of the realism of these prototype tests, usually carried out by the manufacturer's test pilots and without much of the complexity of actual route flying on the airways.* However, certification also requires flights by airline crews who are introduced into the programme at a very early stage, and some hundreds of hours in representative airline route flying but without fare-paying passengers.

On entering airline service, the new aircraft is given over to the two major operating organisations – aircrew and groundcrew – who must then acquire experience of the type. The maintenance personnel learn fault detection, rectification, and procedures for routine servicing, and the removal and replacement of parts at specified times. The reliability of the engines, controls, instruments and structure to the prescribed standards laid down by the designer and implied by the certification for flight is their responsibility, and a major effort is involved in preparing descriptive maintenance handbooks and training several hundreds of engineering personnel in their correct use. In parallel with this the pilots and aircrew gain experience of the new airliner's behaviour in training flights, and learn how to operate, perhaps for the first time, certain new devices or novel controls. Both these operating procedures, on the ground and in the air, are most involved and complex, and errors in either of the learning phases may lead to accidents.

* Cf. Chapter 1 on certification tests, pp. 17–19.

Many of the features of present day civil aircraft have been virtually unchanged for decades, such as the manner of providing altitude control by elevators moved from a control column held in the pilot's hands. But in details each new aircraft is very different from its predecessor, even though some instruments or even the engines may be well known and understood. Furthermore, new personnel are being continuously drafted into the airline staff as others retire, and there are subtle changes over the years in means of instruction and means of carrying out medical and behavioural checks on pilots.

Clearly this inherently fluid situation must be a matter for continuous scrutiny, meticulous attention to detail and the maintenance of rigorous safety standards at all levels. It would be very hard to find fault with the thoroughness of all these stages in the evolution of a new type and its introduction into the total activity of an airline, and in fact, a large proportion of the cost of an aircraft is attributed correctly to maintaining high standards of safety. Yet, even with such care, errors can arise – even if these occur only once in 100 million flights. (See probability chart, p. 113.)

Researchers in this field are now aware that the rare accident frequently results from a series of relatively trivial deviations from normal behaviour, each one of which alone would not necessarily constitute a threat to safety. A combination of such actions, perhaps totalling between four and ten minor errors could, however, lead eventually to total disaster. This basic fact applies to both pilot error in aircraft and 'Captain error' in ships, and, it may seem, can never be entirely eliminated. In these circumstances, it is vital to improve the accuracy of information by which accidents can be correctly explained so that the true nature of the problems may be identified and the correct remedies provided.

Some case histories

To help with the task of identifying problems and providing remedies, it is obviously important to define, or attempt to define, all the elements which may have contributed to a disaster, no matter how trivial some may seem in isolation. The variety of special features in each of the classic air accidents described in the following pages illustrates the complexity of this endeavour.

The accidents discussed here reveal differing degrees of failure in design, foresight, operating procedure, or aircraft handling. It may be instructive to consider whether, in each case, the level of technology of the day could have prevented the situation from arising. Many hard lessons have had to be learned in the course of aeronautical history and it is a fact that some of these accidents arose from inadequate knowledge and experience. In great part, the development of aviation has ensured that many of the fundamental mistakes of the 1930s and 1940s can no longer be repeated; but *other features of these episodes remain to be considered as possible accident causes, even today.*

The Airship R 101

During the 1920s the concept of British Empire air transport received considerable political support. At that time the airship was considered a better contender for longer ranges than the aircraft, and the building of the R 101 was the culmination of many years of research and design in the British Government airship factory. The design of this airship incorporated many novelties, but it was overweight; and to restore the net buoyancy an additional section 77 ft long was incorporated a few weeks before its first demonstration and passenger flight to India. *Only 24 hours were available to flight test this new ship* because of the political timetable set many months before, and most of the crew who flew in this last test flight could not be rested before the passenger flight began on a squally, stormy night in October 1930. Battling through severe gales and rain over the English Channel and with significant loss of gas through chafing of the gas bags, the R 101 pitched downwards in a gust, partially recovered, and finally dived to the earth, catching fire and killing all on board but five.

The detail in which the subsequent enquiry was able to establish certain key facts is of special interest, since this impressive understanding of a complex situation was built up from survivors' stories.*

Thus it became apparent that structural failure or engine deficiency could be discounted as contributory factors; and that there had been no cause for alarm at the time of the change of watch some five minutes before the crash, although a strong wind from the South West was gusting at about 40–50 m.p.h., 'causing the nose of the vessel to move through a considerable angle above and below her horizontal line of flight'.

*This accident took several seconds to take on a serious nature. In contrast, modern aircraft accidents happen extremely rapidly, and the skilled crew, located at the more vulnerable front of the machine, rarely survive. Essential information must therefore be obtained from the crash recorder. Compare the details of the report above with the read-out data given in Fig. 12, p. 128.

It was known that on her trials the R 101 had lost gas through holes worn in the gasbags, and 'perhaps' through her valves as she rolled; and it was stated that on the Indian journey, the ship 'had rolled more than ever before, and had failed to keep height as the Officer of the Watch intended'.

Her height at 2 a.m. when the watch was changed had been 'at least 1000 ft' above the ground and airspeed, a little over 50 knots. The course of the vessel was not directly in the teeth of the wind, and her speed over the ground would have been between 15 and 20 m.p.h.

> ...In these circumstances, at about five minutes past two, her nose dropped and she continued in this position for about 30 seconds, descending rapidly during that period of time. Her pitch downwards was sufficiently severe to wake up a man who was asleep in his bunk, and to cause objects to slide to the lower end of the smoke-room.
>
> The height-coxswain, by putting his elevator up, succeeded at length in bringing the ship again to about an even keel, but she remained in this position only for a few seconds.
>
> ...when it appeared that she was not further responding to up-elevator so as to recover height, the Officer of the Watch gave orders through the engine room telegraph to reduce speed. Orders were also given to release ballast in an effort to lighten the impact.
>
> ...About this moment the vessel got into a second steep dive, which lasted for only a few seconds before she struck the earth. The impact was not severe...(but)...The fire did not break out till after the ship struck the ground.

The question of navigation or piloting error was discussed and it was concluded:

> ...that there is no reason to attribute the accident to any failure in the competence of officers or crew, but that in view of the recent change of watch* and of the prevailing weather it may well have been impossible to bring the ship rapidly back to a horizontal position if her nose was forced down in the way suggested.

It was customary at this period to carry out relatively few flight tests on a new airship – a fact witnessed by the early introduction into service of such craft as the R 100, the Graf Zeppelin and the Hindenburg. But in the case of the R 101, it was not realised that its novel devices should have been proved separately and in combination – and certainly under

---

*A reference to the helmsman's difficulty in getting the 'feel' of the ship.

adverse weather conditions – before a Certificate of Airworthiness was awarded.

It was to be tragically proved that the design, so briefly tested and so hastily certified, was unsafe. Development to more acceptable standards of safety, however, would have involved a period of extensive tests and modifications, in the event denied by the need to meet a wholly non-technical requirement. In consequence, the truth is recognised that no degree of crew skill or anticipation could have overcome the inherent weakness of the R 101. The loss of this airship severely jolted the airworthiness authorities of the UK, and although airship development in Britain was abruptly ended, many of the most brutally acquired lessons remained, and remain today, as baselines in the safety development of heavier-than-air craft.

Night crashes of the Wellington bomber

A disquieting series of accidents to these highly successful twin-engined aircraft occurred during 1940–41 in a period of intensive training of pilots for night flying. All accidents showed similar characteristics, in that about 3 minutes after a normal take-off the aircraft flew into the ground under power, suggesting that the pilot was unaware that his aircraft was losing height. In an eloquent but simple theoretical analysis, Professor A. R. Collar (1949) showed that by day the pilot judged the angle of climb by the force with which he was pressed backwards into his seat, combined with his view of the outside world. At night, however, with little outside vision, the pilot could not distinguish, from the seat pressure alone, the difference between climbing steadily and accelerating.

Thus, after passing from climbing to horizontal flight, the gain in speed would seem like a climb – to which the pilot would respond by pushing the nose of the aircraft down further – in fact, into a fatal dive to the ground.

Another factor in these accidents was represented by the incorporation in this machine of the newly introduced constant-speed propellers. The rotational speed of these would be automatically kept at a pre-set value, irrespective of airspeed or engine power. There would not, therefore, have been any accompanying increase of engine and propeller noise as with the old type of fixed-pitch propeller – to which the pilot would have been accustomed.

It was subsequently realised that pilots should be trained to concentrate on their instruments for unequivocal information regarding speed, rate of climb or descent in darkness. Once this was recognised, and the training altered, the accidents ceased.

The common cause of this series of accidents is recorded as pilot error, but this explanation is correct only insofar as the pilot's action is recognised as but one side of an inevitably fated triangle. The geometry is completed by inadequate flight training, and by failure to anticipate the critical effect on the pilot's perception of the change in propulsion characteristics.

Mid-air collision over the Grand Canyon

On 30 June 1956 two aircraft took off from Los Angeles within 3 minutes of each other, setting off eastwards on initially different routes which later crossed at a distance of about 150 miles. In almost clear conditions above a cloud level of about 20 000 feet the two aircraft, on converging courses, collided virtually at constant speed. The impact tore the tail and rear fuselage from one aeroplane and sliced a major part off the other's wing and aileron, so that both crashed with a total loss of life. These aircraft were travelling according to the permitted regulations outside the supervised air corridors, and although their positions were reported to the ground controllers these were not obliged to advise likelihood of collision.* As originally planned, the aircraft should have had a height separation of 1000 ft; but one pilot had requested an increase of altitude to clear the tops of cloud, and the possible consequences of this had not been recognised.

The forward view from the pilot's cockpit is restricted in the interest both of minimising window area in a pressure cabin and of reducing air resistance; but it is nevertheless surprising that neither the pilots nor other members of the crew saw the other aircraft in time to avoid the collision.

The accident enquiry reported that

> ... the probable cause of this mid-air collision was that the pilots did not see each other in time to avoid the collision. It is not possible to determine why the pilots did not see each other, but the evidence suggests that it resulted from any one or a combination of the following factors:
>
> (1) intervening clouds reducing time for visual separation;
> (2) visual limitations due to cockpit visibility;
> (3) preoccupation with normal cockpit duties;

---

* See Chapter 4, pp. 147–8, 165, on pilot/Air Traffic Control responsibility.

(4) preoccupation with matters unrelated to cockpit duties such as attempting to provide the passengers with a more scenic view of the Grand Canyon area;

(5) physiological limits to human vision reducing the time or opportunity to see and avoid the other aircraft;

(6) insufficiency of en-route air traffic advisory information due to inadequacy of facilities and lack of personnel in air traffic control.

Since there were neither survivors nor eyewitnesses, there is a strong possibility that the collision may indeed have been caused by aircrew error; but what of other possibilities? Passenger or crew illness may have occurred to create those critical moments of distraction. There may have been some major technical failure. The facts may never be resolved, but pilot error remains on the record.

The Elizabethan Crash at Munich – 6 February 1958

The events and aftermath of this accident are described by Captain Bressey in Chapter 1 (pp. 49–53), and discussed by Captain Price in Chapter 6 (pp. 247–8). These contributors offer, respectively, a pilot's view and relevant legal considerations. There is, however, a case for additional comment in a chapter concerned basically with aviation technology. Put simply, this case must rest on the need to protest against a situation in which a pilot can remain effectively penalised in the face of incontrovertible scientific exoneration.

Briefly restated, the facts are that this aircraft made two abortive take-off runs on the slush-covered runway of Munich airport, and on each occasion failed to gain flying speed. A third attempt ended in disaster since the failure to take off – or stop within the available runway distance when the danger was recognised by the pilots – caused the machine to crash through the airport boundary fence and strike a house. As Captain Bressey has remarked, the considerable injury and loss of life which resulted was lent an added poignancy by the popularity and world-status of the passengers, the Manchester United football team: a fact which ensured the perpetuation of this event as an especially painful embarrassment to the British airline concerned.

The conclusion of the German authorities who investigated the accident was that an ice deposit on the upper surface of the wings had resulted in a loss of lift; and had it not been for Captain Thain, this verdict – erroneous as has been shown (pp. 52–3) – would have been allowed to stand. The ten-year struggle for the recognition of Captain Thain's own view – that a layer of slush on the runway increased the drag on the wheels and thereby reduced the acceleration – has also been described in Chapter 1. It is pertinent, however, to note the observation in that chapter, that '... little was known at that time of the retarding effect of slush on the runway ...'

In fact, at that time Canadian airlines appreciated the significance of deep slush as a factor which could dangerously reduce acceleration in the later stages of take-off, for in the winter of 1948–49 the phenomenon had been clearly demonstrated. An aircraft of Trans-Canada Airlines attempting to take-off in heavy, wet slush conditions, failed to accelerate beyond 103 miles per hour, and the pilot found himself unable to lift the nose-wheel clear of the runway surface. A full investigation identified the true effects of slush on drag (see below) and warned pilots that difficulties encountered during attempts to raise the nose-wheel were a positive sign of trouble. Although the Dutch airline KLM similarly instructed their pilots that such a sign justified them in abandoning the take-off, it was not until 1969 that an official British finding confirmed slush as the true cause of the crash at Munich.

It is clear that in the interim the lessons of the Canadian experience had been ignored. For although the claim is recorded that the Canadian report had been circulated to all airlines,* it appears that the implications of that critical precedent in no way inhibited the two Elizabethan pilots from attempting to take off. The conclusion can only be that some *ten years* after the relevant facts had been publicised, these men were unaware of the danger facing them at Munich.

The mathematical inevitability of that danger is illustrated in Fig. 5.

The Trans-Canada Slush Report (Macdonald, 1949) concluded as follows:

It is evident, from the basic figures shown, that the take-off characteristics of the North Star under conditions of wet slush can become very marginal. The added resistance at the wheels has three distinct effects:

(a) The acceleration of the aircraft is severely reduced and thus a greater runway distance is required to attain a safe speed.
(b) Under extreme conditions, the added resistance of the slush (which

---

* Williamson (1972), pp. 114–16. Referring to Mr G. M. Kelly, British accredited representative at the Munich accident investigation, Williamson writes; '... Kelly's copy of the report had arrived with a covering letter from Mr Diament, Chief Engineer of Trans-Canada Airlines, in which he explained that on receiving warning from the research authorities in 1951, about the dangers of slush, he had sent copies of the report to all the major airlines of the world ...'

varies with airspeed) can reach a value at which it nullifies the available thrust, and it prevents further acceleration.

(c) The pitching moment on the nose-wheel builds up very fast and causes a highly unbalanced condition which, if sufficiently great, cannot be overcome by the elevators. The stick force in the cockpit can become excessive in attempting to attain sufficient elevator.

It is recommended that extreme caution be exercised by flight crews when taking off in conditions of wet slush. There are several secondary effects over and above the basic effect as calculated. Such added hazards as iced

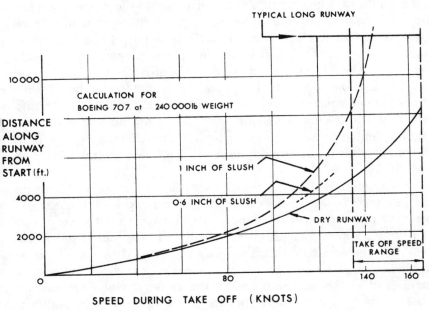

Fig. 5 Effect of slush in increasing runway length requirement

control surfaces and propellers, added weight on the main planes, induction intake blockage and freezing of the wheels in the wheel wells, are all possible to an undetermined degree. It is suggested that take-offs in wet slush be limited to conditions of 2 inches of slush depth. Any indication during the take-off that the nose-wheel cannot be lifted should be regarded as an immediate warning of trouble. The pitching moment on the nose-wheel becomes progressively worse as the speed builds up. Under no conditions should the trim control be used excessively to aid in the movement of the control column as it will cause a severe control hazard if the aircraft does suddenly leave the ground.

Jumbo Jet at San Francisco

Mercifully the 350-seater Boeing 747 has maintained a magnificent record of safety throughout its career. From the date of its first inception into service in December 1969, no less than 226 of these aircraft were on the world's airways by March 1974, logging up a total of 2 million flying hours; a magnificent tribute to its designers and operators.

However, on 30 July 1971 a non-stop flight taking off from San Francisco for Tokyo did not quite reach flying speed when it reached the end of the runway. The under-belly of the aircraft struck threshold lights and a 75 ft long section of a pier projecting into San Francisco Bay; this impact resulted in seven major portions of metal debris piercing the aft fuselage and tail surfaces, injuring passengers and making three out of the four hydraulic systems inoperative. It speaks very well of the inbuilt safety margins of the aircraft design that it was nevertheless able to get into the air and fly around for two hours while the extent of the damage was surveyed from an accompanying aircraft. Although the Boeing veered off the runway during the subsequent landing, and some passengers were injured during egress because of its abnormal angle to the ground, there were otherwise no casualties and no fatalities.

This accident is classic in that it shows how normal safety margins can be eroded by a series of relatively minor adverse events, not one of which alone could have caused the subsequent accident. The crew were expecting a runway length of 9400 ft when in fact only 8500 ft was available. Some part of this deficiency was created by a prohibition forbidding the large Jumbo jets to commence any take-off close to a main road at the landward end of the runway, where jet exhaust blast would have been dangerous for motor car traffic. In addition, there had been changes in wind speed between the time of computation of the take-off and the time when it actually occurred; thus, the actual take-off distance requirement appeared to be some 200–300 ft longer than would be expected under normal prevailing conditions. These encroachments on safety levels were additionally compounded by the pilot's use of 10° of flap during take-off, instead of a recommended 20°; a factor (making for a lower value of lift from the wings) which, at the known weight of the aircraft, pushed the runway requirement out to a wholly unattainable 8675 ft.

That the crew attempted to take off at all under all these circumstances indicates – yet again – a serious breakdown in communications; in fact, subsequent investigation of all the regulations and the data used by the authorities and the crew indicated minor irregularities resulting from failure to amend data books and record the state of runways available at the airport. The ultimate cause of this accident was stated to be the pilot's use of incorrect take-off reference speeds resulting from errors in collection and dissemination of airport information, aircraft despatching, and crew management and discipline. The report published by the Air Transport Safety Board included 21 conclusions and 5 recommendations for avoidance of similar difficulties in future, and 4 recommendations to improve the evacuation of the crew from a grounded aircraft.

The above examples show something of the spectrum which must be covered, and the difficulties which are inherent in determining the cause of aircraft accidents. The problem generated by these complexities may be summed as a dearth of precise information on which it would be possible to make unequivocal judgements. Particularly is this so whenever any question of pilot error arises; and particularly acute is the designer's awareness of the possibility that such errors may be design-induced; in fact, hazards built in by the engineer. Knowledge of this potential must, therefore, be basic to the designer's approach to the creation of an aircraft. The following section outlines some of the areas which offer a special challenge to the designer in the task of establishing a satisfactory and ethical man–machine interface.

Safety by design

The aircraft designer, in striving for safety in a new aircraft, must be very conscious of the history of failures or errors which have been revealed as potential sources of accident. In judging, therefore, how he can modify any aircraft design in the interest of safety, he must decide what is reasonable within normal manufacturing and operating cost limitations, and within the normal behaviour of average pilots in obeying the usual rules of airmanship. Broadly, there are four main spheres of design in which he can exercise this judgement; and in terms of 'designing out' pilot error this means eliminating, insofar as the state of knowledge permits, any engineered situation which implies too high a risk, and any activity which calls for too high a skill, or which demands too great a concentration of pilot effort during involved operational procedures.

Aircraft handling behaviour

Aircraft will get into difficulties if they fly too fast or too slow. The former condition may occur if an autopilot fails and causes the aircraft to nose-dive from high altitude, or may even result from pilot inattention; but such events are rare and are almost invariably remedied since it is a corollary of adequate height that there is time available for correction. In contrast, the stalling of aircraft at low speed – i.e. when the wings are placed at too high an angle to the airflow for the generation of sufficient lift – is much more frequent; and when this condition occurs at any stage during take-off or landing, there is insufficient height to permit recovery.

A typical summary of civil accidents to large transport aircraft over a period of 25 years (ARB, 1971) showed that stalling accidents represented from 7 to 13 per cent of all fatal accidents. The worst stalling accident rates applied to elderly types designed before the Second World War, which fact reflects a growing awareness of the phenomenon, and intensive steps to rectify the situation. Major turbojet transport types in current operation show a stalling accident rate better than 2×10^{-7}, i.e. two accidents occurring in ten million flights.

The designer can, within some limits, extend the angle of incidence before which stalling occurs, but in recent years the demands of high speed have precluded the use of such devices as leading edge slots that can be opened, which were most effective when used many years ago on low-speed types. A solution has been to provide the pilot with a warning of the onset of the stall by means of an electrically-driven control stick shaker, and a further device, a stick pusher, which, if the pilot should overlook the stall warning, automatically moves the stick forward, so reducing the dangerous angle of attack, and leading the aircraft away from the stall (Tomlinson, 1972). Unfortunately, because of the semi-automatic nature of most devices involved in the man–machine relationship, these mechanisms are provided with cut-outs at the command of the pilot for situations in which, either because of equipment malfunction, or a critical close-to-the-ground positioning, the pilot might have good reason for overriding the automatic system.

The designer is aware too, and must make provision against, other potential handling difficulties. Controls may become unstable, or be heavy for the pilot to operate. These characteristics are usually discovered during prototype flying, and can almost always be eliminated – or their effect greatly reduced – by modifications either to the exterior shape of the aircraft, or to the associated systems, before the aircraft enters airline service. Clearly, in difficult situations, the pilot must not be additionally burdened, nor be expected to pay undue attention to matters which are basic engineering responsibilities.

Systems

The aircraft is composed of several systems including the structure, the engine and fuel system, the electrical system, flight controls, cabin air-conditioning, communications system and others. Systems are either designed to have a 'safe life', (i.e. they are tested to many times their actual life by, for example, a factor of 5), or to be 'fail-safe' to an extent that a single failure still leaves the aircraft safe to return to land.

Fail-safe design for the structure is accomplished by providing 'redundant' load paths between two major elements so that a failure occurring in one structural member leads to the redistribution of forces throughout neighbouring members; effectively, this means that even though distorted, the structure can still remain airworthy. For avionic systems – the generic term for aviation electronics – fail-safe characteristics are ensured by duplication, triplication or even quadruplication of units designed so that the failure of one does not prejudice the overall function. In some systems there are 'vote-takers' which monitor the signals from, say, four systems nominally having similar values; should one disagree with the other three it is deduced to be a 'failed' unit and is switched out of circuit.

During the design process the designer employs a technique known as 'failure mode effect analysis', which involves designing a given system in great detail, and deducing all the possible ways in which its individual and collective elements could fail or malfunction. Those findings are related to the total expected flight time of the aircraft, and apportioned to the several segments of the flight, e.g. take-off, climb, cruise, etc., in which the physical conditions imposed on the system – such as temperature, vibration and throughput – vary considerably. By this means potential accident causes can be recognised and corrected by alteration in design, components, or overall method, although in all cases absolute certainty of reliability is impossible, as some of the components to be incorporated in the system are themselves still under development. However, the subsequent object of ground testing before flight is to run the systems for several hours in simulated flight conditions. The analyses of these tests seek for random faults and, later, for 'wear-out' failures. To overcome most of these, avionic systems now have a bench test of about 150 hours, called 'burn-in', which usually identifies particular faults so that after rectification the system settles down to reliable behaviour.

Indicators and controls

All systems need controls and these frequently have distinctly shaped handles or knobs which can be unambiguously recognised even without sight, to minimise the incorrect use of controls. The classic study by Fitts and Jones (1947) analysed 460 pilot error experiences in operating aircraft controls, and identified some forty typical error descrip-

Fig. 6 Distribution of pilot error accident causes 1962–71
In this period, flight crews were held responsible for between 40 and 50 per cent of worldwide fatal air accidents, of which 74 per cent were attributed to errors of airmanship.
(Reproduced from the UK CAA staff newspaper *Airway*, May 1973, and quoted in *Interavia*, p. 105, June 1973)

tions falling into six broad categories. Of these, substitution errors (the wrong control operated) accounted for almost 50 per cent of the total.

The most common of these arose from confusion of engine and throttle controls representing 19 per cent of the total; another 16 per cent were attributed to confusion of flap and wheel controls and 8 per cent in implementing the wrong engine control or feathering buttons. The last 7 per cent under this heading was associated with seventeen other controls. Adjustment and 'forgetting' errors each amounted to a further 18 per cent of all errors. The remaining categories (14 per cent of all errors), included 'reversal' errors, unintentional activation and inability to reach a desired control.

The use of shaped control handles must certainly have gone a long way to improve the situation since that thorough work was published. Although there can be difficulties in operating a control, i.e. it may be so placed that it is difficult to find, or its sense of operation may be unnatural, the solutions – so apparently obvious – may still elude the designer since he can in no way guarantee the behaviour of the user. Dr Allnutt discusses the psychological motivation for departures from 'normal' operating practice in Chapter 2. In the design context, the major problems of the correct use of controls are considered under the headings of work load and cockpit ergonomics. These factors are reviewed in later pages.

Collision avoidance systems (CAS)

The technical problem of an aircraft detecting the presence of another one appearing at a random position with unknown direction and speed remains one of aviation's most difficult hurdles. Airborne radar has proved to be unsuitable because of the high angular accuracy needed; but steps were taken in the late 1960s to develop a world-wide collision-avoidance system. All aircraft using this system would be equipped with an extremely accurate clock controlled by the vibrations of a caesium crystal. Periodically, all aircraft would transmit a radio signal at the same micro-second of time. Coded messages would be received by all other aircraft within range, which would then automatically respond, thus notifying the originating aircraft of a threatening aeroplane's position, speed and direction. An on-board computer would evaluate the collision threat and direct the necessary alteration of course or altitude in order to increase separation distance. However, the logic of this system is not devoid of problems. In common with alternative applications of automation they are discussed in pp. 120–5.

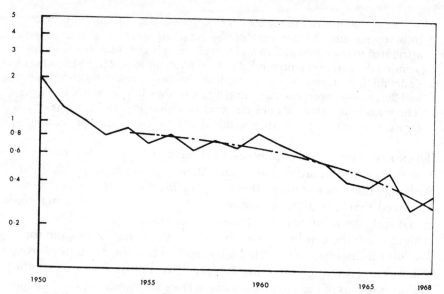

Fig. 7 Passenger fatality rate (Shaw, 1969)
The International Air Transport Association collates all data on world civil aircraft accidents. These are quoted as an accident rate, i.e. number of passengers killed per 100 million passenger kilometres (100 passengers carried for 1000 km = 100 000 passenger km).

The fact that the annual values of passenger accident rate form a smooth, steadily decreasing curve shows that coordinated worldwide efforts to increase safety are taking effect.

The mathematics of safety in design

Accident rate is expressed as either passenger fatalities per 100 million passenger km, or fatal accidents per 100 million aircraft km. The trend of annual plots of such rates indicates the general level of improvement of safety, and a typical curve derived by the International Air Transport Association (IATA) is given in Fig. 7. Such statistics can be analysed further into the proportion of accidents caused by different reasons, as is shown in Fig. 8.

In designing new aircraft and their systems to improve on the previous accident rates, a target figure will be allocated to the aircraft as a whole, e.g. one catastrophic accident in 100 million flights; subsequently a possible source of major risk can be identified and a smaller proportion of the total allocated to this. If there are, for instance, 10 systems which, for the sake of argument, are each assumed to bear an equal risk, then each system must not suffer catastrophic failure in less than 1000 million flights. Further subdivisions within each major system can then be made and a dependable rate evaluated.

In practice many contributory potential failures are not equal and then the question of balancing the improvements must be made. For example, Dr R. Shaw of IATA quotes some typical data in such a case.

	risks per 100 million flights				
CASE	RISK A	RISK B	RISK C	TOTAL	RATE (flights)
1	25	1	100	126	1: 793 000
2	25	5	100	130	1: 769 000
3	25	1	50	76	1:1 317 000

Fig. 8 Additive problems (Shaw, 1969)

He subdivides all flights into three parts and assesses three kinds of risk occurring in different parts of the flight. Case 1, shown in Figure 8, gives a total of 126 corresponding to an accident rate of one in 793 000 hours. If one small risk were increased by a factor of 5 the total is only increased by a small percentage with little effect on the overall result. However, only 50 per cent reduction in the largest risk leads to almost a doubling of the safety level. The lesson from this example is that all risks must be judged relative to each other, and ideally, the best solution will be when they are all equal.

Another way of combining the separate effects Shaw classes as 'factor problems'. This phrase refers to the gain in safety obtained by providing duplicated or triplicated systems, wherein the failure of one would not affect any of the others. Accordingly, the probability of *total* failure is then the product of a probability of failure in each of the individual systems. For instance, if a system fails once in a thousand hours the probability of total failure of all three systems in a one-hour flight is once in 1000 million. Thus, duplication makes the very best of any gains in safety in the individual systems; e.g. in a triplicated system a doubling of reliability improves the overall risk of total failure by a factor of eight.

Although this powerful effect of duplicated systems is mathematically valid, in practical terms the engineer must ensure that there is no possibility of a common error influencing all three systems together. It seems that if all systems are given separate sources of power, controls, transmission cables and indicators, there is virtually no chance of this occurring. However, the human factor introduces its own note of uncertainty; in fact, there have been instances where incorrect maintenance carried out by the same engineer has led to the same fault being embodied on all systems.

The cost of safety

Clearly the provision of duplicated systems and time-consuming ground tests increases costs. Good design and such solutions as specially shaped control handles are negligibly more expensive. Improving the reliability of equipment leads to additional first cost, and to further testing cost, although subsequent in-service maintenance, repair, and replacement costs may be less. It is obvious, however, that an aircraft of high reliability gives confidence to pilots and frees their attention for primary flight duties. Since more faults are likely to occur after the end of a long flight than at its beginning, and since the landing, when the pilots are tired, is the most hazardous part of the flight (Table 1, p. 15), greater unreliability must be a factor in contributing to pilot-error accidents – although the relationship is so involved and, for the pilot, so deep-rooted, that it may be hard to quantify in numerical terms.

The cost of achieving reliability is also reflected in the duration of flight tests. This varies with the size and complexity of the aircraft type and also in response to the ever-increasing number of safety regulations, and the insistence by the certifying authorities on the demonstration of good characteristics.

Although the range of aviation technology is always growing, it is usually directed towards providing higher performance and more sophisticated systems, so that, rather than reducing with time, both flight-test duration and cost show a steady increase. For example, in 1963 the Boeing 727 Trijet, of which over 1000 have now been built, needed a flight testing time of 1137 hours. In 1967 the twin-engined Boeing 737 of much smaller size and weight required almost as much at 1124 hours. By 1969 the far larger 4-engined Boeing 747 Jumbo jet logged a total test time of 1444 hours. (The fact that for these aircraft the flight times are not directly proportionate to their weight is explained by the different numbers of measuring instruments carried on the trials – 3155, 2543 and 5981 respectively.) In comparison, the Concorde flight time on prototype and pre-production aircraft accomplished by 5 February 1975 was 3500 hours out of a planned 4000 hours. The scope of these tests is shown in Table 3.

Table 3 Concorde flight tests

Performance	Field, climb, en route, hold, descend, land, overshoot
Control and handling	All flight modes – engine intake – high angles of attack
Structure	Flutter
Weather conditions	Low temperature ⎱ with high altitude take off High temperature ⎰ Rain Snow and ice – de-icing equipment Crosswinds
Failure representations	Engine failure – single and double Control system malfunctions Instrument standby
System integrity and reliability	Cabin conditioning, auxiliaries, confirmation of ground trials
Airways flying	Communications, radar, displays, automatic flight control system (AFCS) accuracy, performance margins and reserves, compatibility with traffic patterns
Certification	Compliance with international safety regulations on noise

4000 hours total with seven aircraft. Concorde ultimately completed a total of 5335 test hours before receiving the UK Certificate of Airworthiness.

Certainly there is at least one area in which cost has had a positive and lasting effect, for it was recognition of the large costs involved in post-accident investigation which pointed the need to obtain better measured information at the time of an accident. This situation was revolutionised by the gradual introduction of 'black box' flight data recorders designed to survive most crashes, so that such data as height, speed, engine and control settings – recorded prior to the accident – can be played back and analysed. Flight deck recorders are also being introduced. These record all crew conversations and radio messages, so building up a more certain picture of events associated with the flight. Such evidence should go a long way towards removing uncertainties which have in the past been deemed due to pilot error.*

*Boeing's customer support department has a senior accident investigation coordinator whose sole job is to assist the national accident investigation authorities at the scene, and afterwards.

Boeing considers that this staff position improves product quality because it avoids ... 'it crashed' or 'pilot error' accident reports. (*Flight International*, 29 August 1974)

Cockpit workload

The aircraft designer has one type of problem in analysing the behaviour of airflow over a stalling wing, or the reliability of a complex system. He is, however, faced with a totally different and much more intractable kind of problem in attempting to quantify the nature of pilot behaviour in accomplishing complex tasks in the cockpit. During and since the Second World War considerable study has been made of ergonomic factors, the man–machine relationship, and the means of measuring cockpit workload. Unfortunately for the designer – and this point has already been made – the behaviour of even the average pilot cannot be represented mathematically in terms of a uniformly reliable formula. It is true that a considerable number of research experiments have been made with air crews operating in fairly realistic simulated cockpits; but many of the factors under study lend themselves to a variability of behaviour and situation which defy the designer's need for a firm frame of reference.

Nevertheless, mathematical models can offer researchers essential information. An excellent status report of the mathematical representation of pilot behaviour, and the complex problems this work reveals was published by McRuer and Krendel (1974). The variable quantities which affect the pilot's behaviour are indicated in Fig. 9. This diagram shows the relationship between the human pilot and the information with which he is presented and the actions which he must take in controlling the vehicle (aircraft). Although complex, the diagram illustrates very involved relationships which are inherent in evaluating quantitatively the individual relationships given in the lists of variables. Graphic representation of this type is vital to the understanding of such a complex subject; it is then followed by further mathematical relationships covering very many small parts of the total process. One example of these further stages is shown in Fig. 10. This is written in the mathematical notation of control theory and relates the pilot's sensing, processing and muscle responses, all of which are involved as he attempts to deal with unexpected disturbances – for example, wind gusts during a landing manoeuvre. Considering that the relationship of Fig. 10 represents only a small portion of those aspects which have so far been identified by research, the so far ill-defined influences of a psychological nature, and the effect of temporary impairment of sensory or muscular power, the difficulty of producing an acceptable total

mathematical representation of pilot behaviour – as far as the designer
is concerned – is all too clear.

Perhaps the simplest way of identifying the characteristics of pilot

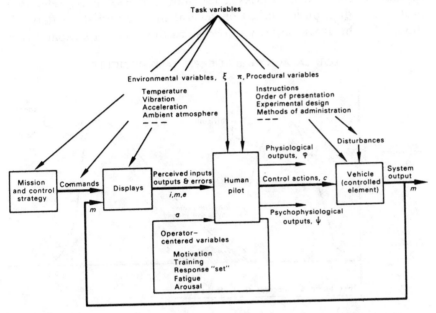

Fig. 9 Variables affecting the pilot/vehicle system (McRuer and Krendel, 1974)

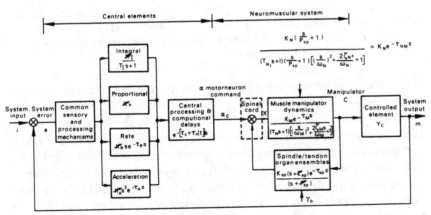

*Fig. 10 Model for human pilot dynamics in single-loop compensatory system with
random-appearing inputs (McRuer and Krendel, 1974)*

workload has been described by Thorne (1973) of Britain's Royal Aircraft Establishment. This method is shown in Fig. 11, wherein pilots' tasks are listed from 'very easy' to 'very hard'. The operator's maximum capacity to perform such tasks is also shown. These curves refer to a probability distribution based on several flights, and show that, although for the great majority of occasions the operator's capability is

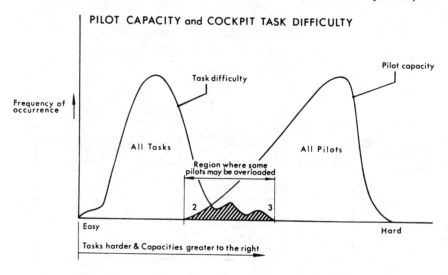

PILOT CAPACITY and COCKPIT TASK DIFFICULTY

Fig. 11 Pilot capacity and cockpit task difficulty (Thorne, 1973)
Task difficulty varies from a few easy items, through many of average difficulty, to a few very difficult items (1).
   The capacity of the majority of pilots lies well above maximum task difficulty most of the time, but below-average pilots· (2) – and even good pilots extremely infrequently (3) – find themselves faced with difficulties beyond their capacity as a result of stress, malfunction of equipment, etc.

well above the level of all difficulties, there are some marginal conditions in which a 'below average' pilot will encounter difficulties beyond him, and even good pilots may be caught out in the very rarest events. Thorne correctly draws attention to the urgent need to reduce the more difficult tasks, and in consequence, to reduce the potential overloading which may occur when many different controls have to be operated in rapid succession during highly accelerating or decelerating flight. Some judgement must be made by the designer of the severity of this congestion of tasks, which will permit him to deploy the cockpit controls and systems so that they can be operated within the range of the pilot's capabilities and limitations. For civil aircraft Thorne concludes that the workload problem can be tackled under five headings:

(a) improvements to the cockpit environment and layout;
(b) altering the nature of the tasks;
(c) reliably automating tasks;
(d) reallocating tasks amongst crew members;
(e) reallocating tasks with respect to time.

The factors which influence pilot behaviour in critical periods of activity in the cockpit include monotony, jet lag, preoccupation with one task at the expense of several concurrent activities, degraded skill in handling emergency situations due to lack of regular practice, and such personal factors as a neurotic disposition and anticipatory tension.

Much research has been devoted to redesigning the whole array of instruments in the cockpit with a view to reducing the frequency with which the pilot must look at different instruments. Early research led to a recognition of typical eye movements of pilots and resulted in a rearrangement of the primary flight controls. A major distraction for the pilot is represented by the necessity for sharing his sighting time between looking outside the cockpit and down at the instruments, since this process is accompanied by a significant change in light level. Captain Bressey (pp. 20–3) has vividly evoked the typical frame for such split concentration in his description of the landing phase. He refers, too, to the fact that military aircraft have taken a lead in providing the head-up display system known as HUD; but other improvements on conventional instrumentation are realistically projected, even though, as will be seen, these are not without their own problems. Instead of having separate dials to indicate speed, height, engine revolutions, etc., many indicators are now combined into one instrument called the flight director. A step beyond this is the use of large cathode ray tubes which can represent not only any of the standard instruments but, in addition, display flight boundaries and desired flight profiles for comparison with the actual situation. The flight director can even plot the position of aircraft in the air traffic control pattern.

While considerable work has been done on these systems they are, unfortunately, very slow to enter into service, not only because of an understandable mistrust of new systems, but also because of the very real necessity for proven stand-by instruments. A further consideration is that the pilot might be severely taxed should he find it necessary to transfer without warning from one kind of presentation to another. Failure of the cathode ray tube itself is not a major hazard, since, in these new arrangements, the tube is duplicated or even triplicated; in the event of failure the system can be switched to another available tube as required.

In this situation the difficulties of the designer in incorporating radically new systems which will prove acceptable to a vast body of regulatory authorities and pilot organisations are formidable. It is nevertheless encouraging to note the great effort which has gone into the development of representative simulation experiments, for where the evaluation of these is sustained by technological fact, essential information is established for the designer.

The range of such tests is typified in the NASA study published in March 1973, wherein a civil aircraft cockpit simulator, complete with controls and displays, and providing the pilot with a visual representation of the outside world and the runway, was driven laterally over a distance of 50 ft to reproduce the motion normally experienced by the pilot in a manoeuvring aircraft. This experiment was used to assess the complex piloting situation generated by the need to overcome the effect of an engine failure during the later stages of take-off. These tests compare jet transports, Jumbo jets and a supersonic type of aircraft; and from the large amount of statistical data obtained by several pilots, safety margins are identified which can then be used as a basis for the mandatory safety features of future designs. However, complex though this representation was, the task was not in fact complicated by pilot workload errors; for the crews were expecting the failure and were not under the normal psychological stresses involved in an actual take-off.* These crews knew they would walk away unharmed from the simulated crash, and it is this type of reservation which so clearly demonstrates the unquantifiable element challenging the engineer.

How automatic?

It is tempting to seek a panacea for pilot error in the increased use of automatic aids to the processes of control in flight. Indeed, tremendous strides in this respect have been made in the last 20 years.

---

*See similar comment by Captain Bressey, on flight manual figures obtained by 'forewarned' test pilots. Chapter 1, pp. 17–19.

With the achievement of automatic blind landing, now in regular service, and sophisticated automatic flight-control systems in which the pilot merely selects the flight regime (e.g. climb, cruise, turn, descent, etc.), there seems to be virtually no limit to the possibilities. However, it remains a regrettable fact that at the very low levels of engineering failure probability which are now current – i.e. one catastrophic failure in ten million hours – the probability of *human* failure is reasonably high. To increase the scope of automatic systems beyond the present standard would lead to significant extra cost and complexity, and more, would require a demonstration of the attainment of improved safety virtually impossible of achievement.

For example, the Boeing 727, as a type, has accumulated over 13 million flying hours producing evidence which has not even reached a 95 per cent probability of demonstrating the one-in-ten-million safety target. Nevertheless, its in-service record is excellent and, in order to make a significant improvement in safety by the incorporation of additional automatic devices, these should have an overall system failure not exceeding 10^{-8} – that is, of one in one hundred million – and this requires the probability of catastrophe from part of the system of the order of 10^{-9}, or one in one thousand million. These low probabilities are difficult to grasp but would be equivalent to one disaster from a single cause during the life of a fleet of 20 000 aircraft operated for 3000 hours each year for 15 years. To demonstrate the achievement of such a safety level would require an accident-free flight-proving many times longer than this; this postulation shows that even if the systems were designed and paid for, the required safety level could not be reasonably demonstrated. This is not to say that steps do not continue to be made towards the improvement and installation of safety devices where these will inhibit pilot error in critical situations.

However, the example quoted by Captain Bressey (pp. 33–4) of the Eastern Airlines Lockheed 1011 which crashed on the approach to Miami airport owing to 'crew preoccupation with a malfunction of the nose landing gear position indicating system' (simply, with the failure of a warning light) points the way for future development in this field. It is clear that the emphasis must lie in the improvement of inherent component and system reliability, and not in the incorporation of an over-abundance of safety measures.

The development of the automatic landing system – popularly known as the 'blind' landing system – commenced in Britain and America in the 1940s. Only 20 years later, systems in the air and on the ground had been proved safe enough for a new aircraft, the Hawker Siddeley Trident, to be designed for operation under the guidance of an auto-landing system. The great advantage in this case lay in the fact that the system was used only for the last few minutes prior to the landing. It could therefore be used on all flights, even in good weather, thus enabling pilots and engineers to gain experience and confidence during many thousands of automatic landings. Further, the ability of the pilot to take over control from the automatic devices during these brief minutes was very certain. The need for blind landing in foggy conditions in Europe is higher than in many other regions and its use throughout a typical year may be seriously required on perhaps only 2 per cent of all flights. The installation of this system may therefore appear to represent a disproportionately large effort for an apparently small part of the operations; but it is judged to be a vital means of saving lives and maintaining the regularity of the scheduled service. It is perhaps ironic that the aircraft manufacturer, as a result, faced a penalty, for the system requires triplexed control lines and many duplicated power supplies. There is a consequent increase in first cost and in running cost; and marginal though this is, it is nevertheless recognised as a disadvantage in selling the aircraft – always in a com-petitive market – in countries which are not troubled by the fog hazard.

As aircraft control systems and the corresponding air traffic control routines and equipment on the ground become more complex and more automatic, so the policy decisions concerned with the future pro-gress of automatic systems become more involved. The development in the mid-1960s of the anti-collision system utilising the caesium crystal clock (described on p. 111) typifies this problem; for the ultimate cost of these high-precision clocks and their associated computation devices – possibly between $30 000 and $50 000 per aircraft – must render such installations quite beyond the reach of the majority of private aircraft operators. The collision risk thus remains, for the private aircraft in a route-conflict situation – unprotected and unheralded by an anti-collision system – presents as lethal a threat as the largest airliner.

An alternative solution which has now been tested by the US Federal Aviation Agency is known as the cooperative type interrogator responder. This presents to the air traffic control on the ground highly accurate radar position data on each individual aircraft, so that possible flight-path conflicts are evaluated in the control centre and pilots are warned accordingly by radio. The changed situation within one decade has resulted from a much higher use of automatic air traffic control displays and more sophisticated computers to predict the future positions of aircraft. Another contemporary factor is the delegation of some of the workload of the ground air traffic controllers to the computer.*

The economics of the new system are also favourable, since at this time an airliner system is expected to cost no more than $10 000, reducing to $1000 for general aircraft. In addition, owing to general improvements in air traffic control procedures, navigational accuracy and awareness of the dangers of mid-air collisions and the conditions likely to lead to them, the numbers of mid-air accidents have been reduced considerably. It is estimated that one-third of the actual collisions, and two-thirds of the fatalities occurring in a recent year might have been avoided by the use of cooperative-type systems.

Although it may be tempting to believe that all transport systems will be completely automated in due course, this assumption is fundamentally misleading. A true understanding of the interaction of automatic systems and the capacity for, and incidence of, human error is elusive because of the low probability factors and the complexities already discussed. Yet significant precedents do exist; the dangers of misjudging the true part which automation can play in public transport systems have been revealed, for instance, in the account of the difficulties which beset another major concept, the San Francisco BARTD Underground system.

This undertaking was proposed in the 1960s as a modern, 75 mile long, 80 m.p.h. commuter railway connecting many outlying areas with the central business district. On the basis of the high level of automation successfully employed in manned spaceflight, it was suggested that the system would be fully automatic, the trains starting and stopping in stations automatically, with only some external signal box control for route-changing at points. An economic advantage was to be gained by eliminating the payroll of train crews. In the event, however, it was found that, in spite of considerable care in design and testing, safety levels could not be raised to the extremely high requirements of the public transport system, the crux of the problem being that it would have been necessary (but difficult, if not impossible) for the variety of statutory bodies authorising the system to produce evidence to safeguard that authorisation. Complete redesign of the system for greater integrity, or the alternative of a lengthy develop-

---

* See typical computer application, p. 153.

Table 4 Man–machine capabilities (NASA, 1968)

Human superiority	Machine superiority
1. Originality (ability to arrive at new, different problem solutions)	1. Precise, repetitive operations
2. Reprogramming rapidly (as in acquiring new procedures)	2. Reacting with minimum lag (in microseconds, not milliseconds)
3. Recognising certain types of impending failures quickly (by sensing changes in mechanical and acoustic vibrations)	3. Storing and recalling large amounts of data
4. Detecting signals (as radar scope returns) in high-noise environments	4. Sensitivity to stimuli (machines sense energy in bands beyond man's sensitivity spectrum)
5. Performing and operating though task-overloaded	5. Monitoring functions (even under stress conditions)
6. Providing a logical description of events (to amplify, clarify, negate other data)	6. Exerting large amounts of force
7. Reasoning inductively (in diagnosing a general condition from specific symptoms)	7. Reasoning deductively (in identifying a specific item as belonging to a larger class)
8. Handling unexpected occurrences (as in evaluating alternate risks and selecting the optimal alternate, or corrective action)	
9. Utilising equipment beyond its limits as necessary (i.e., advantageously using equipment factors of safety)	

One of man's greatest limitations when functioning as a system component is his low information-handling rate, even when he is engaged in a single task. This limitation, in turn, is further degraded by man's limited buffer storage (immediate memory) capacity. On the plus side, however, are:

(1) Man's ability to handle a great variety of different information-processing tasks
(2) Man's capability to adapt to new tasks or environments and learn new skills
(3) Man's judgemental ability in devising newly required procedures or resolving unexpected contingencies

The primary objective, therefore, in man–machine analysis is not the determination of whether man will do a better job than a machine but rather of whether he can do an *adequate* one for less money, less weight, and less power and with a smaller probability of failure and need for maintenance.

ment and correction process, would have been prohibitively expensive; alterations were therefore made to permit *manual* driving with supervision to correct for defects and system failures.

This was indeed a bold attempt to make a decisive advance towards a fully automatic system. Yet the manner in which the project became progressively diluted by the need to overcome immediate practical difficulties serves as a salutary reminder that the fully automatic public transport system is still a very remote possibility.

It is the designer's belief that the true place of the sophisticated control system and computer is certainly not as a substitute for man as controller and decision maker in charge of a public transport system. These installations and devices will always be slaves, perhaps more and more intelligent and helpful, and more dependable than in the past; and they will certainly off-load many of the routine tasks associated with the complex piloting of present-day aircraft. Far from replacing Man, however, their use will better enable him to devote his unique capability to those areas of decision in which only he can perform.

To provide guidelines for the designers of manned spacecraft systems in judging which parts of the piloting process should be made automatic, and which should be left firmly under the pilot's direct control, a comparison of the capabilities of man and machine was published in 1968 by the National Aeronautics and Space Administration of the USA (Table 4).

Feedback

Feedback* is the generic term for the records of crashes, the statistical derivations from large numbers of such records, and the analysis of individual crashes, with the action subsequently taken. Documentation of this nature offers a foundation for the accident investigator, for in any accident all possible sources of error are likely at the outset. Similar methods may therefore be used in obtaining information concerning failures of equipment, or concerning flying errors; but, for reasons which will become apparent, the essential historical record on

---

* From cybernetics (the study of control and communication mechanisms in machines and in living creatures: *Chambers's 20th Century Dictionary*). Feedback involves detecting an error in a system and feeding a correction to this error back to that part of the system capable of applying the corrective action: thus, the error is subsequently reduced to zero.

pilot error is mainly distinguished for the gaps which exist in the chain of information.

For every accident there are many more 'incidents'. It has long been recognised that reporting of incidents and the accumulation of the evidence from these must give a better picture of potential accidents, thus enabling precautions to be taken in due time.* Dr Roxbee Cox made this point as far back as 1949. Yet the difficulties in setting up a world-wide reporting system to monitor these eventualities have until recently defied resolution for many reasons. Dr Allnutt examines some of the psychological motivations more fully in Chapter 2, but, in essence, pilots, like most human beings, may feel themselves embarrassed by any incident which may reflect on their pride, or their personal competence – or which, if revealed, might imperil, or might be thought to imperil, their job security. The common reaction, then, is to withhold information, doubtless with the determination that the incident will alert them to the potential difficulty and prevent them falling into a similar situation in the future.†

Airlines, too, jealous of their reputations for safety, are hardly likely to enthuse over the publication of large numbers of such incidents which may weaken public confidence in their operations. However, steps were taken in 1973 to set up a confidential procedure which would enable incidents to be reported in an agreed and consistent manner, and scrutinised by independent authorities. Applied to the behaviour of particular aircraft employed in several airlines, this wider assessment would, it is felt, be of more value than the limited experience obtained by any one of them in isolation. The independent assessors would make recommendations to those authorities who have the means of correction within their power – such as the certifying authorities, the airline, or the aircraft manufacturer. These data do not become public and are not available to the parties contending the legal proceedings which may follow an accident.‡

In the technical analysis of accidents a great deal of special expertise was developed during the Second World War. There were several thousand crashes during that period, and techniques of reconstruction

---

* This point is discussed in some detail in Chapter 4, pp. 151–2.
† 'But we do not live in an ideal world. We all make everyday judgements of what we think is "important" or "trivial". We observe the ties of friendship or loyalty in deciding whether to tell third parties about a colleague's error. We are sensitive about our own error partly because the word "error" itself has an emotive content: it signifies a fall from grace – and partly because there can be repercussions in terms of self-esteem or career.' (Caplan.) See also p. 88.
‡ But see Chapter 6, pp. 247–8, for lawyers' reliance on the experts' evidence.

evolved were based on such methods as predicting the sequence of break-up in the air from the trail of accident wreckage, the likely direction and height of the aircraft, its speed, and that of the wind. Microscopic examination of damage to engines, structure and instruments can differentiate surprisingly well between primary damage occurring in the air and secondary damage resulting from the subsequent ground impact. These techniques are helpful in deciding whether a control may have broken, or may have jammed, whether engines were operating or shut down, and whether the aircraft was in a fully flying condition or in a stalled state. In many cases such evidence could point decisively to the cause of accident, and the question of pilot error could not arise. In too many cases, however, research cannot ascertain whether a control was incorrectly operated or whether it had failed; and, if the former, why the pilot had done this, and whether it was an erroneous act or not.

It is for such reasons, and with the aim of generally improving the standard of accident analysis, that the black box crash recorder became a mandatory fitment of aircraft in the 1960s.

The device can certainly offer some bearing on the question of pilot error; but at a time of critically high workload in the cockpit, and disorientation caused by a multiplicity of failures, even the black box recorder cannot provide all the necessary data. Hence, in order to further improve knowledge of the cockpit situation, crew voice recorders are now being advocated, and will become increasingly operational in the late 1970s. It is hoped that this equipment will go far towards eliminating the uncertainty which surrounds the assessment of pilot-error responsibility, particularly in those instances where the crew are in a position to discuss situations of difficulty or danger prior to an actual event.*

The final analysis of many accidents is a complex matter, and the more so if pilot error is thought to be a likely cause. It is, of course, this fact which largely accounts for the lag (shown in Table 2, p. 57) between aircraft accidents and the date of the final accident report. A year or more may also elapse before alternative 'cures' are designed, flight tested and finally introduced into the many hundreds of the aircraft type which are already in service. The improvement of aircraft design in recognition of the findings of aircraft accident investigations

---

*See Prologue, Chapter 4, p. 145 for report of transcript of conversation from cockpit recorder recovered after TWA B727 crash.

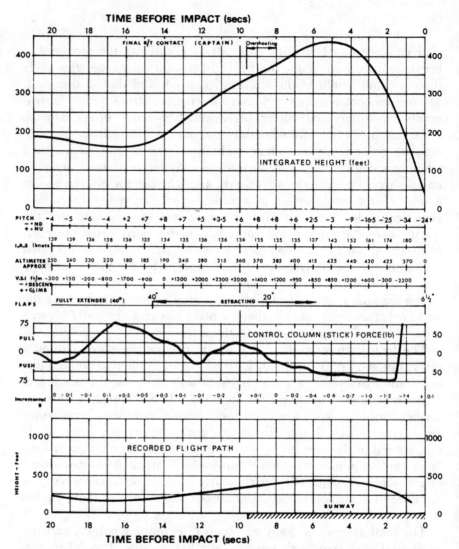

Fig. 12 Coordinated flight data (by permission of HMSO)

Read-out from black box flight-recorder recovered from BEA Vanguard G-APEE which crashed at night in misty conditions at London Airport, 27 October 1965.

The top graph shows the height above the runway. Starting from the left, there is a descent towards the runway, which is then abandoned, because of poor visibility, at about 17 seconds before the crash. Here the aircraft first levels out in response to the pull of the control column (shown in the middle graph); then follows a steep climb which is sensed by the pilot who, in response, moves the control column forward at about 12 seconds. However, the climb continues as speed falls away to the low value of 135 knots at 7 seconds. After this the aircraft appears to be pushed into a dive and the pilot applies a sharp correcting pull of the control column only one second before impact.

Also shown are values of pitch angle, vertical speed and altimeter height.

is a continuous process, and such lessons find their way into the national standards of airworthiness. The British Civil Aircraft Requirements (BCAR), for example, contain no fewer than 10 000 items of guidance which must be followed in the design and subsequent testing of a new type. Bearing in mind the complexity of an aircraft and the very high levels of safety achieved, this does not follow without a very considerable amount of persevering effort from the many hundreds of people involved in designing and making large modern aircraft.

Future prospects

It is pertinent here to discuss changes for the future, firstly in the short term, and then in the longer view. It is equally relevant to consider the influences which may determine the course of events and increase the risk of pilot-error episodes in the context of a given airline (i.e. one of substance) operating a well-known type of aircraft over known routes.

Without doubt one such factor lies in the increased sensitivity of the public to aircraft noise within the noise envelope of the airport and on the busy routes traversed. To alleviate this nuisance, airliners now climb more steeply after take-off, thereafter reducing engine power and climbing more slowly; but there is evidence that this practice has reduced desirable power/safety margins, and in fact was considered as a possible contributory factor in the Trident crash at Staines, England, in 1972 (Chapter 1, pp. 60–1). Among the negative influences must also be numbered the increase in air traffic, particularly at peak hours, and at holiday airports. This is reflected in the inordinately large number of aircraft waiting in the stacks – a matter of concern and increased workload for both pilot and air traffic controller.

It is not yet possible to gauge fully the long-term effect of the world fuel shortage, save that the need to conserve more expensive fuel has been responsible for alterations in operational flight plans which may, in some special circumstances, reduce fuel margins. There is an additional possibility of crew errors in judging fuel allowances in unfavourable combinations of weather, and unexpected traffic delays. Further risks are represented by unusual, nationwide failures; a repetition of the total loss of electric power, for instance, which occurred in the north eastern USA in 1964, might lead to some loss of communications, information and landing aids, in spite of generally good stand-by systems; and here again, any or all of these contingencies would put an unexpected workload on flight crews.

Finally, the outbreak of politically inspired, criminal, or simply maniacal hijacking and bombing activities in aircraft in flight has subjected aircrews to substantially higher anxiety, fatigue and personal risk, all of which may increase the prospect of pilot error as seen in classical terms; that is, as against the behavioural standards which could be expected to prevail given normal circumstances. It must be said, however, that the safety record of air crews in such circumstances is entirely a matter for admiration.

It is logical, next, to attempt to define some of the areas, in both procedure and design, in which changes could be made to ameliorate factors conducive to pilot error. As witnessed in Chapter 1 (pp. 27–33), fatigue plays its part in the deterioration of pilot behaviour in critical situations, and a basic move must be the universal acceptance of intelligently constructed duty schedules which fully compensate for the effort and stress involved. It is possible, too, that some improvement may be gained from more rigorous physical and psychological tests – from stricter selection, coupled with more active training for emergency procedures. The undue number of accidents which occur during training flights may call for the further use of simulators,* but these could not be employed exclusively, since there remains the problem of the 'false' element common to analogous rather than actual situations. However, it may be argued that the incidence of the pilot-error syndrome could be appreciably reduced were more generous margins to be allocated for the whole of the piloting task, in all phases of flight.† For example, if engineering design could ensure that take-off were to become possible within a smaller proportion of the runway length, more time would be available for critical decisions in the event of engine failure at speed. Accidents such as that at San Francisco (see p. 106) or Munich (see p. 49) would be eliminated.

Other improvements might be made in increasing engine power, or in the speed of control response; but it has been shown that these would impose extensive penalties in terms of the aircraft's overall size, cost or complexity. Since, to be truly valid, such improvements would necessitate changes in many other features, the combined effect would

* *Daily Telegraph*, 7 June 1974.
† See workload/flight-time and safety analysis, Chapter 5, pp. 182–4.

lead to a massive increase in aircraft cost. Such an escalation would inevitably be reflected, not only throughout the aviation industry, but additionally, as an adverse factor in the national economy. The proposal to provide larger numbers of crew members in order to share tasks and thus reduce workload must be treated with similar caution. Larger crews must mean a more complex pattern of task-sharing and delegation; and at critical times, only the Captain can make the vital decision.*

For the longer-term future there are many possibilities, and some new ideas which must be classed as speculative. Although the prospect for some novelties appears to exist, the successful outcome of these would depend on considerable and lengthy research to answer fundamental questions, practical engineering adoption of the schemes, and proof that the primary advantages envisaged were not accompanied by unexpected secondary disadvantages. Such novel ideas can be illustrated by three examples.

Firstly, it has been proposed that as an aircraft proceeds through the scheming, designing and testing phases, a comprehensive simulator should grow in parallel with the design, so that from the earliest possible moment a complete representation of the cockpit and all its ergonomic characteristics can be assessed. This would ensure a far better feedback between the designers, the operators and the safety authorities and so make for higher factors of safety. Fewer aircraft losses would well repay the cost and the technological effort associated with this additional activity.

Secondly, there are exciting possibilities stemming from the understanding of the man–machine relationship which is now a subject of considerable study and experiment in aircraft, spacecraft and other systems. Pilots already employ the senses of sight, sound and touch; suggestions have been made that other relationships could be made between the aircraft and the more basic features of the complex human perceptual systems. The human nerve system transmits information round the body at a speed of 50–100 metres per second, at from 10 to more than 100 pulses per second, to as many as half-a-million motor units throughout the body. As more is learned of the behaviour of the neural system combined with known and standard ways of moving fingers, limbs and head, perhaps better and more comprehensive rela-

* Cf. this approach to the crew-complement problem with that of Captain Leibing, Chapter 5, pp. 174–91.

tionships could be evolved between the pilot and the highly complex information outputs from the aircraft. Finesse of touch and control reside in the finger tips; and experimental aircraft have flown in which the major flight control elements are finger operated from special controls mounted at the side rather than in front of the pilot.

Similarly, the brain incorporates highly sophisticated spatial and angular velocity detectors which might be more directly related to gyroscopic and other sensors in the aircraft. It appears that a very great revolution in what has been called the 'man–machine interface' might be possible in the future, which could improve the rate and certainty of sensory inputs to the pilot. If this concept can be successfully realised, it must raise his capacity well above the threshold demanded by the piloting task, even in hitherto overload situations.

The third possibility embodies a totally automatic supervisory monitor for the whole piloting function. The idea is essentially derived from the practice adopted over a century ago in the railways, in which errors by the signalman operating manual levers to move signals and points are prevented by a mechanical interlock frame which physically prevents incorrect actions. No such aeronautical equivalent could possibly be defined during the period when piloting involved a pilot in a seat merely looking at instrument dials in front of him; but today, a totally different situation exists.

Virtually all the quantities involved in the piloting process now appear in electronic data form to feed the automatic pilot system, the automatic landing system, electronic engine control system, black box crash recorder, engine health monitoring unit, and in some cases, the collision avoidance system. All these devices have emerged from separate developments, each designed to improve a small part of the total operating process.

The next step towards the concept of the totally automatic monitor is to combine all the electronic outputs within a new unified system. In this would be encoded the idealised sequence of controlling events, both in quantity and time, corresponding to the best operation of the aircraft in relation to its design and the conditions for which it was certified. It is certainly not intended that the pilot should be removed from the controlling loop; rather, he is still in the seat performing his traditional function in the same way as the signalman in his box. His every action, however, is monitored second by second and compared by the electronic measurements with the idealised programme built into the unit. Any significant departure from the plan could either lead to visual or aural warning signals or to the imposition of control-movement restraints which he would detect as an opposition to the control action he was proposing to take.

Such a technique seems a logical next step in the integration of the spasmodically derived individual automatic parts in the aircraft; but in practical terms it would require the production of a totally new system to replace all the existing parts. In terms of economics this is indeed a gigantic step which may be difficult enough to take. It would, in any case, require a revolutionary approach of similar magnitude. However, if it can be demonstrated by design and experiment that such a total monitor could increase safety levels by factors of 10 or 100, then the undoubtedly great expense involved in producing the system would be repaid many times by the saving of life and resources over several years.

It is interesting, too, to consider the possible emergence of totally new classes of aircraft. Concorde – the supersonic transport designed and built by Britain and France – is now in commercial service and, in spite of its novel shape, aerodynamics and flight profile, has proved itself to be no more difficult to handle than any other in the collection of well-behaved aircraft that travel the airways. There is certainly no foundation for the suggestion that its pilots might better be found from the ranks of the supersonic fighter pilots, than from those airline captains used to dealing with the more sedate subsonic types. Although the scale of the flying functions is very different, the general behaviour of an SST is really very little different from that of the whole army of aircraft on which all airline safety has been logged; and hence, in piloting terms, no really radical departure from previous practice is to be expected.

Even faster transport systems have been proposed, going back as far as 1933, when Dr Eugen Sänger published designs for a rocket-propelled winged aircraft capable of flying around the world. In recent years, however, US designers have originated studies, and structural, propulsion, and aerodynamic experiments, on hypersonic aircraft which would cruise at speeds of from 5–10 times the speed of sound. This type of air transport would be appropriate to intercontinental flights of many thousands of miles, but would be extremely costly to produce. Yet, even more startling designs have been envisaged in which passengers would fly in rocket-propelled ballistic vehicles derived from the application of spaceflight techniques. At the present time such possibilities remain either as speculation, or as experiments, since in any

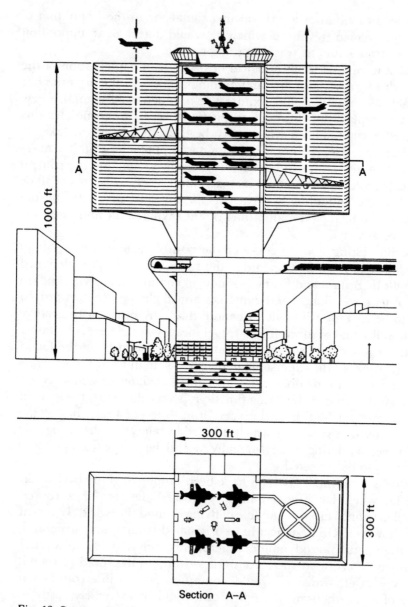

Fig. 13 Compact city airport
Features: small ground area; lifts for multiple access; noise shielding; adequate space for waiting aircraft; interport tube rail service.

case, their translation into practical systems would require the invest-
ment of national resources on a scale which must surpass even that of
the bi-national Concorde effort.

More immediate possibilities, however, are suggested by the development of V/STOL (Vertical/Short Take-Off and Landing) transport
aircraft. Considerable effort went into this project in the 1960s following the successful demonstration of vertical jet flight by aircraft such
as the Hawker Siddeley Harrier, which could also fly at supersonic
speeds. It is therefore quite probable that, in the future, such airliners
will land and take off vertically along a beam from a city airport under
fully automatic control. The pilot would concentrate on checking the
serviceability of all engines and systems; but otherwise, what have
hitherto been the most dangerous phases of normal flight would be
delegated to the automatic operation. Apart from the obvious advantages associated with this new system, i.e. the capacity to enter a metropolitan region with little noise, the airport, occupying no more space
than a normal office block, would clearly revolutionise the safety aspects
of civil transport. The arrangement proposed is shown in Fig. 13.

Although reference so far has been made to manned aircraft, there
are fascinating parallels to be found in the realms of manned spaceflight. Here totally different levels of safety are involved with 'flights'
lasting several days, in which mechanical failure and pilot error could
be disastrous. The associated systems were therefore designed with considerable margins of redundancy so that failed systems could be by-
passed or reversionary modes adopted. Throughout these spaceflights
engineering functions are telemetered to earth where teams of special-
ists study, second by second, the values of the physical qualities pertaining in the spacecraft, even during its sojourn on the lunar surface.

It is obvious, of course, that such a complex and expensive remote
monitoring system designed for relatively few spaceflights could not be
applied to the multiple flights of airliners on the airways. Nevertheless
it was proposed as early as 1951 that some on-ground monitoring of
engineering functions in airliners could be done by radio links. The
suggestion is an interesting example of the information 'spin-off' which
has been generated by the space programme, and it is without doubt
that future phases of this programme will contribute equally significantly to commercial air transportation. Valuable lessons then, can be
expected from the development and use of the 'space-shuttle' – the US
'piggy-back' orbital aeroplane which represents a fascinating half-way
house between the aircraft and the spacecraft.

The pilot would use a cockpit instrument panel of broadly conventional type, and would have manual control, with the ability to override and alter automatic functions. Once again, these would be special flights, and the technique of ground monitoring is likely to be employed.

The existence of all the above possibilities offers a bewildering range of choice to the scientist and the engineer, and it may be rash to predict the course of future action in respect of aviation development. However, there are signs that it may be possible to resolve some of these uncertainties in a way which must lead to the production and operation of vastly safer aircraft in the future.

It is now recognised that the world no longer is of infinite size or possesses infinite resources, and movements are afoot from many different areas towards what are termed 'invariant' transport systems. These will be orientated towards the needs of a more or less stable world population, and based on the likely distribution of people between new towns, big cities and the rural areas. The concept will call for the identification of a relatively small number of transportation systems appropriate to the classes of travel distances;* e.g. within the city, inter-city, inter-country, and transcontinental. All these systems will be designed to face the environmental problems of noise, pollution, space-conflict and energy. They will not depend on fuels possessing a recognised exhaustion point, but will preferably be related to the invariant energy systems such as those derived from solar power, and energy obtained from the sea. As such systems are developed, those responsible for design will be faced with the need to establish approaches to the problems of safety in keeping with the social and technological advances inherent in such changes. They will, of course, be aided by the existence of a solid body of knowledge bequeathed by generations of experience in the operation of highly sophisticated transportation systems; but of particular significance will be the fact that this knowledge, and the current state of the art, will be applied

---

* The energy crisis initiated by the Middle East oil-producing countries after the Arab–Israeli Yom-Kippur War of October 1973, resulted in a serious scientific approach to 'low-technology' transport.

One such project at the University of Oxford's Department of Engineering Science began in 1974 with financial support from OXFAM, and concerned itself with the development of manual-powered transport – such as a 'modernised' wheelbarrow, and bicycle-type trucks and water-transport. These vehicles were designed for the economy and environment of the Third World. Other projects included a pedal-driven corn-grinder, a winch, a soya bean flour-miller, and a pedal-driven maize sheller.

outside the constraints which influence and pose problems for contemporary transport operations.

In this chapter it has been necessary to discuss the wider aspects of safety in aviation, in order to identify the part played by pilot error. The span of this review must now narrow once more in order to focus on the sad, and seemingly incessant, reiteration of this phrase; and to do so it is appropriate to conclude with two quotations.

The first of these is taken from a newspaper report* and should be placed in its historical perspective. It appears in the seventh decade of aviation; almost 30 years after the publication of the pilot-error studies made by Fitts and Jones and some 16 years after the ill-judged verdict on the Munich air disaster. The finding of 'pilot error' against the Captain and co-pilot of a chartered Vanguard aircraft which crashed in a snowstorm, near Basle, Switzerland, in 1973, killing the two pilots and 108 passengers – is described thus:

> '... The crash was caused by faulty reception of navigational radio signals. This resulted in a pilot's error,' a Swiss inquiry said ... 'The imperfect reception was due to atmospheric disturbances, but it may be been compounded by defective equipment aboard the plane.'

The second quotation is made from a paper read by Dr R. R. Shaw of IATA to the Society of Licensed Aircraft Engineers and Technologists (Shaw 1969).

> Some sixty per cent of all accidents involve major factors which can be dismissed as 'pilot error'. This sort of diagnosis gives a ... feeling of self-righteousness to those who work on the ground; but I want to state categorically that I do not believe in pilot error as a major cause of accidents. There are, it is true, a very few rare cases where it seems clear that the pilot wilfully ignored proper procedures and got himself into a situation which led to an accident. But this sort of thing perhaps accounts for one or two per cent of accidents – not sixty per cent. Apart from the rare exceptions, pilots are a ... responsible professional group of men, and pilot error accidents occur, not because they have been sloppy, careless, or wilfully disobedient, but *because we on the ground have laid booby traps for them, into which they have finally fallen.*

---

* *Daily Telegraph*, 7 June 1974. The British Accident Report (No. 11/75) was published in September 1975.

References

AGARD (1968) *Problems of Cockpit Environment*. Conference Proceedings No. 55, AGARD Advisory Group for Aerospace Research and Development. Amsterdam.

AGARD (1970) *Education and Training in Aerospace Medicine*. AGARD-CP-75-70.

AGARD (1972) *Automation in Manned Aerospace Systems*. Conference Proceedings No. 114. NATO.

AGARD (1973a) *The Use of Nystagmography in Aviation Medicine*. AGARD-CP-128.

AGARD (1973b) *Behavioural Aspects of Aircraft Accidents*. AGARD-CP-132.

AGARD (1973c) *Clinical Psychology and Psychiatry of the Aerospace Operational Environment*. AGARD-CP-133.

ARB (1971a) *A Review of Accidents Caused by Failure of Flying Control Systems from 1947 to 1968 Inclusive*. Technical Note No. 102, Issue 1, June.

ARB (1971b) *Transport Aircraft Stalling Accident Rates*. Technical Note No. 103, Issue 1, September.

Association of European Airlines (1974) *Air Transportation in Europe*.

Australia, Dept of Civil Aviation (1972) *Aviation Safety Digest* (81), September.

Aviation Week and Space Technology (1971a) Board probing 747 takeoff accident. 9 August, 26.

Aviation Week and Space Technology (1971b) NTSB broadens 747 investigation. 16 August, 29.

Aviation Week and Space Technology (1972a) Takeoff data cited in 747 accident. 30 October, 63.

Aviation Week and Space Technology (1972b) NTSB elucidates cause of 747 mishap. 6 November, 53.

Aviation Week and Space Technology (1973a) F-86 crash cause. 23 April, 67.

Aviation Week and Space Technology (1973b) ICAO details Libyan airliner's last flight. 16 July, 85.

Aviation Week and Space Technology (1973c) NTSB analyses O'Hare runway accident. 1 October, 65.

Aviation Week and Space Technology (1974) Soviets tighten air safety rules: serious shortcomings admitted. 14 January, 26.

Barnaby, K. C. (1968) *Some Ship Disasters and their Causes*. Hutchinson, London.

Benson, A. J. (Ed.) (1974) *Orientation/Disorientation Training of Flying Personnel: A Working Group Report*. AGARD-R-625.

Benson, A. J. and Burchard, E. (1973) *Spatial Disorientation in Flight. A Handbook for Aircrew*. AGARDograph No. 170, NATO.

Bergin (1949) Accidents, Chapter XXXIV in *Aviation Medicine*. John Wright, Bristol.

Bloom, J. N. (Ed.) (1974) *Principles of Avionics. Computer Systems*. AGARD-AG-183.

Caplan, H. (1955) The investigation of aircraft accidents and incidents. *Journal of the Royal Aeronautical Society*, **59,** January 1955, 45.

Caplan, H. (1972) *Safety Information Systems*. Paper No. 11 in technical symposium 'Outlook on Safety', British Airline Pilots Association.

Chapanis, A. (1965) *Man–Machine Engineering*. Tavistock, London.

Civil Aviation Authority (1972) *Accidents to Aircraft on the British Register, 1971*. CAP 361, HMSO, London.

Civil Aviation Authority (1973a) *Britain's Civil Aviation Authority*. CAA, London.

Civil Aviation Authority (1973b) *Report of the Committee on Flight Time Limitations*. CAA, London.

Civil Aviation Authority (1973c) *Trident I.G.-ARP1. Report on Public Enquiry into the Causes and Circumstances of the Accident near Staines on 18 June 1972*. Chapter V, *Conclusions*. HMSO, London.

Civil Aviation Authority (1975) *Airmiss Analysis Report*. Press notice, 3 February.

Colegate, R. (1973) The Civil Aviation Act 1971 – a general view. Symposium on Regulation of British Aviation in the 1970s. *Royal Aeronautical Society Journal*, **77** (753).

Collar, A. R. (1949) *On an Aspect of the Accidental History of Aircraft Taking Off at Night*. Reports and Memoranda No. 2277 (9872), HMSO, London.

Coombs, L. F. E. (1972a) Harrow, 20 years on. *Modern Railways*, October, 374.

Coombs, L. F. E. (1972b) Visual perception and high-speed trains. *Modern Railways*, December, 462.

Coombs, L. F. E. (1973a) Flight deck evolution – Part 1. *Interavia* (7), 774.

Coombs, L. F. E. (1973b) Flight deck evolution – Part 2. *Interavia* (8), 879.

Coombs, L. F. E. (1974) *Left and Right in Cockpit Evolution*. Paper presented to the Historical Group of the Royal Aeronautical Society. Smiths Industries Ltd, London.

Cowin, H. W. (1974) Integrated electronic displays. *Flight International*, **105,** 3 January, 19.

Davis, R. D. (1948) *Pilot Error*. Air Ministry AP 3139A. HMSO, London.

Elliott Brothers (London) Ltd (1971) *Automatic Flight Control System for SST Concorde: Elliott-Sfena.*

Engineering (1968) The vital 'black box'. 23 August, 287.

Eula, E. and Capodagli, G. (1973) Engine condition monitoring – the Alitalia approach. *Shell Aviation News*, **420,** 24.

Field, H. (1974) The experience gap. *Flight International*, **105,** 30 May, 696.

Fitts, P. M. and Jones, R. E. (1947) *Analysis of Factors Contributing to 460 'Pilot-Error' experiences in Operating Aircraft Controls*. Mem. Rep. TSEAA-694-12, 1 July. *Psychological Aspects of Instrument Display: 1. Analysis of 270 'Pilot-Error' Experiences in Reading and Interpreting Aircraft Instruments*. Mem. Rep. TSEAA-694-12A, 1 October. Aero Medical Laboratory, Air Material Command, Wright-Patterson Air Force Base, Dayton, Ohio.

Flight International (1973a) British airline safety. **103,** 12 April, 573.

Flight International (1973b) Air safety priorities. Editorial, **103,** 19 April.

Flight International (1973c) Air transport: Trident crash findings. **103,** 10 May, 694.

Flight International (1973d) Accident prevention. Editorial, **103,** 17 May.

Flight International (1973e) One in a million flights. Editorial, **104,** 9 August.

Flight International (1974a) Air transport: 'reckless practices' should be deterred, says BCAL. **105,** 14 February, 190.

Flight International (1974b) Systems No. 22: Is standby electrical power really necessary? **105,** 7 March, 295.

Flight International (1974c) Concorde: BAC replies to Mr Benn. **105,** 4 April, 426.

Flight Safety Foundation Inc. (1973) *Flight Safety Facts and Analysis*, **4**(1) and (2).

Flight Safety Foundation Inc. (1975) Taxonomy of pilot error factor. *Human Factors Bulletin* (Arlington, Virginia), January/February.

Fort, T. E. (1974) Avionics No. 180: Flight recorders. *Flight International*, **105,** 30 May, 713.

Francis, B. (1973) CRT displays promote flight-deck safety. *Design Engineering*, May, 53.

Frisby, C. B. (1947) Field research in flying training. *Occupational Psychology*, January.

Frost, J. D. Jr. (1974) *Fabrication of Neurophysiological Monitoring Systems.* NASA CR-2416.

Gilson, C. (1973) Air safety: after a military aircraft accident. *Flight International*, **104,** 8 November, 781.

HMSO (1931) *Report of the R101 Enquiry. Part VI: Discussion of Cause of Disaster.* 3825, HMSO, London.

Hornsby, M. (1973) Inquiry finds pilot error caused Delhi crash. *The Times*, 21 June.

Hutchinson, J. E. (1974) The approach hazard. *Shell Aviation News*, **426,** 10.

ICAO (1957) *Douglas DC7, N 6324 Collided over the Grand Canyon, Arizona on 30 June 1956.* US Civil Aeronautics Board Accident Investigation Report SA-320, File No. 1–0090, 17 April. Circular 54-AN/49 (22), 95.

ICAO (1968a) Why mid-air collisions? *Aircraft Accident Digest* (17), Circular 88-AN/74, **1,** 175.

ICAO (1968b) The pilot's role in preventing mid-air collisions. *Aircraft Accident Digest* (17), Circular 88-AN/74, **I,** 180.

ICAO (1969) *Aircraft Accident Digest* (17), Circular 88-AN/74, **III.**

ICAO (1971) *Aircraft Accident Digest* (18), Circular 96-AN/79, **III.**

Interavia Air Letter (1973a) Reflections on problems facing flight crews. No. 7, 705.

Interavia Air Letter (1973b) Need for Aerosat. No. 7783, 20 June.

Interavia Air Letter (1973c) Aircraft incident and defect reporting to be mandatory. No. 7788, 27 June, 4.

Interavia Air Letter (1974a) Aircraft losses in 1973. No. 7919, 9 January, 2.

Interavia Air Letter (1974b) NTSB recommends new flight recorder legislation. No. 7969, 20 March, 4.

Interavia Air Letter (1974c) UK aircraft to carry cockpit voice recorders. No. 8018, 31 May, 2.

Interavia Air Letter (1975) FAA tightens air safety procedures. No. 8164, 7 January, 5.

International Civil Aviation Organisation (1959) *Manual of Aircraft Accident Investigation*, 3rd edn. Doc 6920-AN/855/3.

Johnson, L. C. and Naitch, P. (1974) *The Operational Consequences of Sleep Deprivation and Sleep Deficit.* AGARDograph No. 193.

Klass, P. J. (1973) ILS monitor techniques show promise. *Aviation Week and Space Technology,* 13 August, 22.

Klass, P. J. (1974) Anti-collision systems report readied. *Aviation Week and Space Technology,* 11 February, 38.

Kowalsky, N. B. *et al.* (1974) *An Analysis of Pilot-Error-Related Aircraft Accidents.* NASA Contractor Report CR-2444. Washington, DC.

Kruszewski, E. T. and Thomson, R. G. (1963) *Development of Airframe Design Technology for Crashworthiness.* SAE Report No. 730319, April.

Lucking, A. J. (1973) Industrial relations and the Trident disaster. *Flight International,* **104,** 19 July, 88.

Macdonald, I. S. (1949) *The Effect of Slush on the Take-off Characteristics of the North Star Aircraft.* Engineering Report R0405-4-1, Trans-Canada Airlines, Winnipeg, Manitoba.

McRuer, D. T. and Krendel, E. S. (1974) *Mathematical Models of Human Pilot Behaviour.* AGARD-AG-188.

Mead, P. H. (1973) Reliability in avionics. *Flight International,* **104,** 29 November, 900.

Morgan, W. P. (1973) the first 2000 flying hours (Concorde). *Esso Air World,* **26** (2), 42.

Morgan, W. P. (1973) Total simulation – a near future goal. *Shell Aviation News,* **420,** 8.

Moseley, H. G. (1961) Thirty aircraft accidents, in *Aerospace Medicine,* H. G. Armstrong, Ed., Baillière, Tindall, London.

Mowray, G. H. and Gebhard, J. W. (1958) Man's senses as information channels. Paper 5 in *Selected Papers on Human Failures.* Johns Hopkins University Applied Physics Laboratory, Maryland.

NASA (1968) *An Introduction to the Assurance of Human Performance in Space Systems.* SP-6506.

Nicholson, Tony (1973) Work, rest and safety in the air. *New Scientist,* 17 May, 404.

Owen, J. B. B. and Grinsted, F. (1949) *The Investigation of Aircraft Accidents Involving Airframe Failure.* Reports and Memoranda No. 2300, HMSO, London.

Pilot (1974) Pilot safety: Accidents only happen to other pilots. December, 18.

Ramsden, J. M. (1973) World airline safety: a 'flight' analysis. *Flight International,* **103,** 17 May, 737.

Ramsden, J. M. (1974) Air safety: the uncommon cause. *Flight International*, **105,** 30 May, 691.

Ramsden, J. M. (1974) The safe airline. Part I: The manufacturer. *Flight International*, **106,** 29 August, Supplement 33–36.

Ramsden, J. M. (1974) Airline safety reporting. *Flight International*, **106,** 12 September.

Ramsden, J. M. (1974) The safe airline. Part II: Government regulator. *Flight International*, **106,** 28 November, 761–4.

Ramsden, J. M. (1974) Human factors. *Flight International*, **106,** 28 November, 764–8.

Ramsden, J. M. (1974) Towards an overall safety index? *Flight International*, **106,** 28 November, 768–9.

Ramsden, J. M. (1974) 1974 world air transport safety record. *Flight International*, **106,** 28 November, 769–70.

Ramsden, J. M. (1974) General aviation safety. *Flight International*, **106,** 28 November, 771.

Ramsden, J. M. (1974) The UK military safety record. *Flight International*, **106,** 28 November, 772–3.

Ramsden, J. M. (1975) Sensitive safety sources. *Flight International*, **2** (2), 1.

Roes, A. (1972) *Flight Safety Aerodynamics*, 2nd edn. Roed, Linköping, Sweden.

Roscoe, S. N. (1974) *Man as a Precious Resource: The Enhancement of Human Effectiveness in Air Transport Operations*. Aviation Research Laboratory, Institute of Aviation, Illinois.

Roscoe, S. N. and Kraus, E. F. (1973) Pilotage error and residual attention: the evaluation of a performance control system in airborne area navigation. *Navigation*, **20** (3).

Roscoe, S. N., Williges, R. C. and Hopkins, C. O. (1972) The new aviation scientist – psychologist and engineer. Innovations in training (P. W. Clement, Ed.). *Professional Psychology*, summer.

Royal Aeronautical Society Air Law Group (1973) *International Aircraft Accidents Investigation*. Symposium, 15 January.

Scano, A. (1974) *Survey of Current Cardiovascular and Respiratory Examination Methods in Medical Selection and Control of Aircrew*. AGARD-AG-196.

Shaw, R. R. (1969) *Safety – Man and Machine*. Second Sholto Douglas Lecture, SLAET, London, 9 July.

Shepherd, E. C. (1973) Braving the weather aloft. *New Scientist*, 8 March, 554.

Sherif, M. (1952) The consequences of elimination of stable anchorages in individual and group situations. *Sociometry* (University of Oklahoma), **XV**, 272–305.

Snyder, C. T., Drinkwater, F. J. III and Fry, E. B. (1973) *Takeoff Certification Considerations for Large Subsonic and Supersonic Transport Airplanes using the Ames Flight Simulator for Advanced Aircraft.* TN D-7106, NASA, March.

Spanner, E. F. (1931) *The Tragedy of 'R 101'*, vols I and II. E. F. Spanner, London.

Sperry Systems (1973) *Multifunction and Other Displays.* Pub. No. 81–0678–01–05, Phoenix, Arizona.

Sturgeon, J. R. (1972) *Influence of Pilot and Aircraft Characteristics on Structural Loads in Operational Flight.* AGARD-R-608, NATO, September.

Thorne, R. G. (1973) *Research in Human Engineering at the RAE.* Technical Memorandum EP 558, RAE.

Tomlinson, B. N. (1972) *Direct Lift Control in a Large Transport Aircraft – a Simulator Study of Proportional DLC.* RAE T.R. 72154.

Tulpule, A. H. (1974) *An Analysis of Some World Transport Statistics.* Transport and Road Research Laboratory Report No. 622.

Tye, W. (1975) Airworthiness – yesterday, today, tomorrow? *Aerospace*, **2** (3), 12.

Williamson, S. (1972) *The Munich Air Disaster.* Cassirer, Plymouth, Devon.

Williamson, S. (1973) Disasters and designs: The Munich air disaster. *Economist*, 31 March, 94.

Wright, P. (1974) Biology: hazards in fast time-zone shifts. *The Times* science report, 1 February.

4

Air Traffic Control Factors
Philip Martin

Prologue

From *The Times*, London, 29 January 1975, datelined Washington, 28 January.

Several weeks before the crash on December 1 of a Trans World Airlines (TWA) Boeing 727 west of Washington, another big jet airliner came perilously close to hitting the same mountain range.

According to Government and airline sources, officials of the second airline, said to have been United, warned the Federal Aviation Administration (FAA) that there was dangerous confusion over traffic instructions issued for landings at Dulles International Airport ... about 20 miles west of the capital. An American Airlines pilot told (the Inquiry) how he had received a landing clearance similar to TWA's just an hour before the accident. But because his interpretation of the rules was different from that made by the TWA crew, he specifically asked what altitude he should maintain, and was safely guided by controllers all the way on to the Dulles runway.

The TWA crew's fatal reading of the rules was demonstrated in a transcript of conversation from the crash-resistant cockpit recorder recovered from the wreck.

'You know', the captain said, 'according to this dumb sheet it says thirty-four hundred (3400) to Round Hill is our minimum altitude.' The sheet referred to was the approach chart showing the standard procedure for an approach to Dulles runway. But, after some discussion with the other two cockpit crewmen, the pilot concluded that the rules allowed an immediate descent to 1800 feet because the traffic controller had radioed that the aircraft was cleared for its approach.

... Even an 1800 feet minimum would have allowed the jet to skim just above the

1764 feet ridge. But the pressure effects of the fierce winds whipping westwards across the peak are believed to have accounted for the extra loss of altitude.

This incident is merely one among many such unhappy episodes. Because these cases reflect so many negative factors in the pilot/controller dialogue, and because, too often, one of these factors may be the apportionment of 'blame', it is necessary for the enquirer to juxtapose the declared rules and objectives of air traffic control with the everyday realities of that service. The rules, if not immutable, are nevertheless positively stated. The frame in which they are interpreted and executed, however, – or as in the above case, *not* executed – differs with the participant's viewpoint. For the pilot any 'difficulties' represent frustrations measured against an ascending scale of danger. For the controller, the 'difficulties' are the constraints enforced in any attempt to manipulate limited human and technical resources within a finite system. Not least among these constraints is the controller's knowledge of the penalties of carelessness or error; but let the platform from which he contends be described.

1. The airspace environment

From the moment that an aeroplane set out to fly from *A* to *B* at the same time as another aeroplane set out to fly from *B* to *A*, it became necessary to devise some means of ensuring that they did not collide en route, especially if that route was enshrouded in cloud. This situation, of course, became more complex as air traffic increased, air routes proliferated over topographical hazards and major cities alike, and aircraft performance improved. Thus, Air Traffic Control (ATC) was born.

ATC is a ground-based service dedicated (in ICAO parlance) to the achievement of a safe, orderly and expeditious movement of air traffic. Its essential aim is to prevent collision in the air by providing instructions, guidance and advice to pilots through the medium of radiotelephony (R/T) voice communications. It can be said, therefore, that where ATC is functioning efficiently (as it usually does) the possibility of accidents and incidents occurring due to pilot error, which involve other aircraft, should be very small.

Why, then, is the pilot held responsible, from time to time, for errors in an ATC environment?

There are three main reasons for this: firstly, the unsatisfactory nature of the basic rules of the air; secondly, the inadequacies of the ATC service itself; and, thirdly, the inherent division of responsibility between the pilot and the air traffic controllers for the safety of the aircraft. Let us look at each of these.

The Rules of the Air are laid down by ICAO in an Annex (No. 2) to the Chicago Convention on International Civil Aviation, which took place in 1944. They have been modified in detail from time to time but, essentially, they recognise and legalise two types of flight operation, namely, that in which the pilot is solely responsible for the avoidance of other aircraft and that in which ATC assumes that responsibility. The international rules in the first of these cases are the Visual Flight Rules (VFR) and, in the second, the Instrument Flight Rules (IFR). The Visual Flight Rules have become generally known by the descriptive term 'see and be seen'.

Now the VFR are only applicable if the aircraft flies in weather conditions within certain prescribed limits, known as Visual Meteorological Conditions. These limits are set out in Table 5, taken from ICAO Annex 2.

As will be seen, these limits vary according to whether the aircraft

Table 5 Visual meteorological conditions

	Within controlled airspace		Outside controlled airspace	
	Above	At or below	Above	At or below
	900 m (3000 ft) above mean sea level or 300 m (1000 ft) above terrain, whichever is higher*			
Flight visibility	8 km (5 miles)	8 km (5 miles) (5 km (3 miles)+	8 km (5 miles)	1·5 km (1 mile)‡
Distance from cloud: (a) horizontal (b) vertical	1·5 km (1 mile) 300 m (1000 ft)	1·5 km (1 mile) 300 m (1000 ft)	1·5 km (1 mile) 300 m (1000 ft)	Clear of clouds and in sight of the ground or water

*Unless a higher plane of division is prescribed on the basis of regional air navigation agreements or by the appropriate ATS (Air Traffic Services) authority.
+When so prescribed by the appropriate ATS authority.
‡Except that helicopters may operate with a flight visibility below 1·5 km (1 mile) if manœuvred at a speed that gives adequate opportunity to observe other traffic or any obstructions in time to avoid collision.

is operating within or outside controlled airspace (of which more will follow) and whether it is operating above or below specified heights; but, basically, they are: flight visibility, 5 miles; distance from cloud, 1 mile horizontally and 1000 ft vertically. Flights are thus not permitted to operate under these rules at night nor above flight level 200 (i.e. about 20 000 ft) without special authorisation.

The IFR apply whenever the flying conditions are worse than those defined for visual meteorological conditions. These rules require, among other things, that the aircraft shall carry suitable instruments and navigational and communications equipment to enable it to operate in these flying conditions, known appropriately as Instrument Meteorological Conditions (IMC). Normally, all IFR flights are also required to submit a flight plan, report position regularly and conform to specified altitudes or flight levels.

The other important distinction is between 'controlled' and 'uncontrolled' airspace. Controlled airspace *is that airspace within which air traffic control service is provided*; the converse applies to uncontrolled airspace. There are several forms of controlled airspace; some applicable to IFR flights only, some to IFR and controlled VFR flights only, and some within which VFR flights are permitted, but are not subject to control. Even in uncontrolled airspace, ATC normally provides a flight information or advisory service.

This, then, is the airspace environment in which the pilot of today flies his aircraft. If he is a private pilot, not licensed for IFR flight, nor his aeroplane suitably equipped, he conforms to VFR. These rules give him a considerable amount of freedom to fly how and when he wishes, provided he assumes the responsibility of avoiding other aircraft while he is doing so. If, on the other hand, he is a commercial transport pilot, he endeavours to fly at all times under IFR and in controlled airspace. He does this because he seeks, above all, the protection from other aircraft which, under IFR in controlled airspace, he can obtain from ATC. He is, in any case, severely restricted in his ability to keep a lookout for other aircraft on account of the limited vision obtainable from the window of the modern jet aircraft cockpit and because closing speeds are so high – two jets approaching head-on en route would do so at around 1000 knots – that it is just not practical anyway. However, a major problem arises when the two types of operation, VFR and IFR, are being carried out in the same airspace which, it will be recalled, is permissible in one of the categories of controlled airspace. This is a situation which results, with melancholy frequency, in fatal accidents for which one or other, or both, of the unfortunate pilots concerned are assigned the blame.

A typical accident of the kind occurred in September 1969, when an Allegheny Airlines, Indianapolis, DC 9 and a Forth Corporation Piper PA28, collided in flight approximately 4 miles northwest of Fairland, Indiana, USA. All 82 occupants, 78 passengers and 4 crew members aboard the DC 9 and the pilot of the PA 28 were fatally injured. Both aircraft were destroyed by the collision and ground impact. The Allegheny DC 9 was under positive radar control of the US Federal Aviation Administration's Indianapolis Approach Control, descending from 6000 ft to an assigned altitude of 2500 ft at the time of the collision. The Piper PA 28 was being flown by a student pilot on a solo cross-country run in accordance with a VFR flight plan. The collision occurred at an altitude of approximately 3550 ft. The visibility in the area was at least 15 miles, but there was an intervening cloud condition which precluded the crew of either aircraft from sighting the other until a few seconds prior to the collision.

The probable cause of this accident was determined to be the deficiencies in the collision avoidance capability of the ATC system of the Federal Aviation Administration in a terminal area in which mixed IFR and VFR traffic operated. The deficiencies included the inadequacy of the see-and-avoid concept under the circumstances of this case, the technical limitations of radar in detecting all aircraft, and the absence of Federal Aviation Regulations which would provide a system of adequate separation of mixed VFR and IFR traffic in terminal areas.*

It should be noted that this accident took place in an ATC environment. The underlying fault with accidents of this kind lies not with ATC, nor with the pilots, but with the Rules of the Air. Ideally, of course, *all* flights should be under control, but a moment's reflection will show that the provision of all the necessary ground and airborne facilities to achieve this universally would be prohibitively expensive and would severely restrict the freedom of movement of the non-transport aircraft population. In the USA, where the problem has assumed critical proportions, electronic airborne proximity warning and collision avoidance devices have been developed for installation in, at least, the aircraft of the commercial airlines. At the same time, moves are afoot to reconfigure the airspace in some busy traffic areas so that IFR and VFR flights in these areas are, as far as possible,

* From the *World Airline Accident Summary* (ARB), 9 October 1969.

segregated. This, together with mandatory controlled operation under IFR at specified high-density locations where there are mixed transport and non-transport operations, seems to be the logical solution to the problem.

The foregoing discussion leads naturally to the second ATC aspect of pilot error – the inadequacies of the ATC system. They are many; and most of them are a function of technological deficiencies and economic considerations.

An ATC system, in fact, is just one element of the overall Air Traffic Services (ATS) infrastructure, comprising the facilities and services necessary to enable aircraft to operate in a safe and orderly manner. These facilities and services include (in addition to ATC) navigational aids, communication channels, meteorological services, lighting, search and rescue, and aeronautical information services. ATC, in essence, consists of the personnel, buildings and equipment by means of which traffic movement can be controlled. The essential link between ATC and the aircraft, as previously mentioned, is radiotelephony.

The quality of ATC varies enormously throughout the world. In many places, there is none at all. In some busy areas, such as high-density terminal areas in North America and Western Europe, ATC is quite sophisticated and is dependent to a large extent on automated equipment and techniques. Generally, the type of ATC service given is related to the nature and volume of the traffic offering.

Basically, there are two methods of controlling air traffic. These are usually referred to as 'procedural' and 'radar'. Procedural control involves the application of the lateral, longitudinal and vertical separation standards laid down by ICAO on the basis of position and height information supplied by the aircraft. As these separation standards are quite large (for instance, 10 minutes longitudinally and 1000 ft vertically), they are not suitable for the expeditious movement of traffic in high density areas. In such areas, radar is employed as the basic ATC separation tool, and this permits the use of horizontal minima of the order of 3–5 nautical miles. In practice, many ATC units use a mixture of procedural and radar control, applying procedural separation vertically and radar separation horizontally. Where it is not practicable to use radar, such as in the North Atlantic and Pacific oceanic areas, it is necessary for ATC to apply procedural control exclusively.

2. The Controller

It is true that ATC occasionally fails both pilot and controller. It does so because the complexities and deficiencies described not only compound errors but also critically, and sometimes disastrously, reduce the time available for their correction. In ideal circumstances any analysis of these problems would depend on an important basis of detailed historical reports covering the not inconsiderable number of incidents which have occurred, and on an 'on-line' system of reporting capable of updating the record. In reality, however, a high proportion of errors within the ATC environment are never reported, for there is a great degree of sympathetic appreciation of the mutual problems of pilot and controller. It is customary as a result, for errors arising during heavy, complex or difficult traffic situations to be dismissed by a brief apology, and it is, in fact, only when emotions are stirred by an actual or potential hazard that an official report is made. This is human, but unfortunate, since statistical information derived from such a limited reporting procedure may be not merely valueless, but positively misleading.

There can be no doubt whatsoever that a full reporting system would indicate areas of common fault in flight and control techniques, and the need for appropriate remedies; but the establishment of such a reporting system, desirable though it may be, is by no means a simple matter. Because errors frequently occur during busy periods, the controller is in no position to make copious notes at such times, but must wait until he is relieved at the end of his duty period. Inevitably, other events and activities have supervened, and the clarity of the controller's memory has suffered accordingly. Thus, the 'body of knowledge' essential to the improvement of any technology is seriously diminished, and, until this weakness is corrected, the consequences of the 'information gaps' in the pilot/controller interface must – and will – continue to assert themselves.

Mandatory reporting of incidents has been introduced by a few national authorities* but is not an international procedure. The success or otherwise of such a system will depend upon the degree of cooperation obtained from pilots and controllers in reporting all incidents, no matter how trivial they may appear at the time. The cooperation between pilots and controllers is such that it is always abhorrent for one

* See pp. 126, 239–40.

to report on the other; however it must be appreciated that while errors can be repeated on many occasions without causing a serious incident, their timely identification would enable appropriate changes – aimed at the elimination of such errors – to be made in procedures.

Control decisions and the ideal system

Pilots and controllers are required to make many decisions affecting aircraft safety and expedition. The high speed of aircraft operating in all planes (speed/rate of climb/descent/change of direction)...demands that such decisions must be made quickly and accurately. The requirement for 'safe' decisions is, of course, paramount, since the speed of operations, and the numbers and complexity of aircraft in relatively small sectors of airspace, may well preclude a wrong decision from being corrected in time to avoid the development of a potentially hazardous situation. Decisions made under the pressures of fast-moving and momentous events may not always be the best or most efficient; but delay – or the total lack of decision – can lead to embarrassing complications causing increased workload and a lowering of the overall level of safety.

Decisions in these circumstances can only be safe provided they require the minimum of consideration in 'thinking time' and 'calculation'. They must, therefore, follow simple and constantly practised patterns, and in general should be accepted variations of a known, pre-planned pattern of operation within a relatively simple, standardised system. The necessity for simplicity and standardisation is, of course, a constantly recurring theme in all aspects of air operations, and is in particular, keenly felt by the pilot who operates in many different national and international spheres.

It is not difficult to set out the basic requirements for safe and efficient operations within the ATC environment. Standardisation and simplicity, of course; the use of uni-directional tracks to minimise closing speeds; the avoidance of ambiguity; and the provision of adequate time – time for concentration; time for the assimilation of information and for considered decision ... and time to carry out the required tasks. The discussion which follows is concerned with those aspects of current ATC practice which fail to satisfy these requirements. Each element of failure represents its special difficulties for the controller; but transmuted by the urgencies of the airborne operation, they become in turn, problems for the pilot.

3. Communications

For all its efficiency, radio telephone (R/T) communication remains a somewhat tenuous thread between pilot and controller since, for various reasons, the link may be broken. In practice, this is not common, and such breaks normally occur only for short periods before contact is re-established. Unfortunately, ATC instructions frequently can only ensure safe separation for relatively short distances ahead of the aircraft, and any break in the R/T link – or hiatus in pilot/controller communication – can therefore be potentially serious. In this context major problems can arise from any, or all, of a number of causes.

The influence of weather effects, for instance, may be reflected in excessive interference, normally occurring during thunderstorms or heavy precipitation. R/T loading is frequently high on these occasions due to the increased workload generated by these effects; but such difficulties are obviously natural hazards. Other burdens on the ATC service, however, are negative attributes of the communications system, and would benefit from an urgent and determined appraisal aimed at their eradication. On the simplest level, ATC locations suffer from the lack of any capability for readily available and easily handled permanent record, and pilots and controllers are far too often faced with the necessity to repeat information. A solution to this problem has been outlined earlier, and indeed must lie in the direction of the development – and general incorporation – of the 'automatic data interchange', whereby pilot and controller are freed from their critical reliance on R/T. Instead, such information as flight and meteorological data will be presented to them on tape, or in the form of computerised print-out.

Meanwhile, the universal dependence on R/T continues to throw up its special and disturbing challenges, the most basic of which stems from the use of non-standard phraseology, and the consequent risk of misunderstanding generated by 'random' conversation. Given a communications system which depends heavily on standardised and universally accepted patterns of phrase and response, the importance of strict R/T discipline and adherence to such forms is plain. In this context it is necessary to emphasise the difficulties experienced by international flying and ATC personnel who may have only the most rudimentary grasp of English (hopefully alleged to be the 'international language' of aviation). In the UK, the courses at the major centres of training for the Commercial Pilot Licence and Air Traffic Controller ratings in-corporate English language tuition for all foreign students. Coupled with the need for intensive professional training, however, the course time scales – some 54 weeks in the case of the pilot-trainees, and 2–4 months for ATC students on typical Aerodrome Approach and/or Radar courses – leave no room for anything but the acquisition of the most basic capacity in English; although, of course, all practical and theoretical training is conducted in that language. There appears to be a clear onus on international airline operators and civil aviation authorities to ensure that aircrew and ATC employees – who may well have decades of service ahead of them – achieve an appropriate com-petence in English. Any deficiency in this respect must remain as a continuing source of hazard.*

The use of numbers in R/T communications is particularly sensitive to error, since airways radio frequencies are commonly saturated by number groups. These may represent aircraft trip numbers, flight levels, barometric data, runways in use, headings and distances, and a variety of equally important numerically coded instructions. Un-fortunately, numbers are inevitably duplicated and at any one time

---

*An example is on record. The following paragraphs are quoted from a letter in *Aviation Week* and Space Technology, 12 January 1976. The author, Vice Admiral Allen M. Shinn USN(Ret.) served on the Navy Court of Inquiry into the events described:

> On Feb 25, 1960 a Navy DC6 transport collided with a Brazilian (Real Aero Lineas) DC3 over Rio de Janeiro harbour, killing 61 persons.
> ...at the time of the collision, and for some 5 minutes before, both aircraft were under IFR and under direct voice control by Rio approach control. The same controller made all trans-missions to both aircraft, using the same...frequency for both *but using fluent Portuguese for communication with the Real DC3 and* English for the Navy DC6...(*using a phrase book*) for the latter
> ...all evidence...proved conclusively that use of two different languages to conduct Rio air traffic control was a major cause – if not the primary cause – of the accident. This made it impossible for the two plane crews – neither of which spoke the other's language – to know that the controller had given them clearances which led them to the same point in the sky...

Vice Admiral Shinn's letter also deprecated the current language impasse in Canada, where moves are afoot to establish French as an 'official' language of the Quebec ATC system, in deference to ethnic sentiment. His letter ends:

> Use of language always has deep political overtones. Whether French or English should be used at Quebec in the end (will) probably be decided by politicians who have no sense of air safety, but only a sense of votes on the ground. But...if Quebec, or any other modern air traffic control system, attempts to use two or more languages simultaneously, a replay of the fatal Rio collision is most certainly lurking in the wings.

Cf. this allusion to political influence on safety factors with Captain Leibing's thesis in Chapter 5. See especially pp. 177–9.

the controller may find two aircraft operated by different companies on his frequency, both using the same 'flight number', or two aircraft with *similar* 'flight numbers' but with a single company prefix: e.g., Air India 112 departs Heathrow at 1136 followed by Royal Jordanian 112 at 1139 both following the same route; or North East 441 reports over Brecon at 0738, North East 401 reports over Brecon at 0740. The potential for error is obvious.

Number confusion can be responsible for the selection of an incorrect R/T frequency, such as 132.05 instead of 132.45. This risk is particularly acute where a number of similar frequencies are used in one area, and may result in a temporary loss of contact. The consequences of this may be aggravated by the current lack of an immediate alternative 'loss of contact' frequency or other method of emergency communication – an omission which can only be regarded as a serious weakness in the system. Pilots *do* select incorrect frequencies, and in some cases make contact with another position within the same ATC unit. On identification – necessitating a check with other duty controllers – the caller is transferred to the correct frequency with only a short delay, albeit he has increased the workload of another controller unnecessarily. However, at times the pilot either makes contact with another unit, which is hard pressed to assist him, or he may make no contact at all, in which case he spends time checking his equipment, trying his alternative equipment, and finally returning to his original frequency – if he remembers it.

This situation can cause significant delay in establishing contact with the appropriate controller, but, together with other problems associated with congested R/T frequencies, could be alleviated by the allocation of one additional radio frequency at Area Control Centres. The extra frequency should be sited adjacent to a permanently occupied position (perhaps at the very nerve centre of the Operations Room, the supervisor's suite) and under normal traffic conditions would not require to be continuously manned, since an automatic alert would indicate any call from an aircraft. The adoption of this system, of course, would ensure that any aircraft losing contact would have a standard frequency available, provided for just such a contingency.

Under abnormal conditions of heavy traffic delays, however, the additional location would be manned and operated as an information frequency for aircraft operating within the Controlled Airspace system; this service would offer information on delays, diversion airfields, and updated weather conditions, etc.

The extra frequency represents a further gain; there is no doubt that, under current operational conditions, pilots listening to a busy R/T frequency, and appreciating the controller's workload, hesitate to request additional information which may not be imperative to the safety of their aircraft but might well assist and improve the efficient operation of their flight. From the controller's point of view it would be a tremendous advantage if, during such periods of heavy traffic, he had no requirement to obtain and pass flight information but could merely refer aircraft to an alternative frequency.

In sum, although R/T is the primary medium of ATC communications, it must be agreed that the qualitatively primitive nature of current speech techniques leaves significant room for improvement. Mutual and essential understanding between pilot and controller, for instance, implies accurate reception and full comprehension of the messages being passed; but under the conditions of the terminal area environment there is, simply, insufficient time to read all messages back – although this is done for those considered to be of the highest importance.

It has already been stated that one of the basic requirements for safe operation is time for the assimilation of the information received. The admission that the scale and pace of air traffic movements often precludes this indicates the fundamental risk in this sphere of air operations.

It is not difficult to choose a typical illustration of this risk, for it will doubtless be remembered that among the human weaknesses explored by Dr Allnutt in Chapter 2 (p. 74) is the phenomenon that it is perfectly possible for the human being to hear what he expects to hear – particularly if he is being distracted by other considerations at the time. In the ATC experience, this propensity may be encouraged by the existence of such factors as similar beacon identifications; an aircraft requesting a route via 'Biggin Hill', for instance, may be cleared by the controller via 'Beacon Hill'. The pilot, failing to note the difference, proceeds in the wrong direction; and thus adds a second link to a potential chain of errors.*

4. Route structure and navigation

Analysis of these two elements exposes a number of deficiencies in their present method of operation and, with equal clarity, the reasons for

_____
*Following such an incident in the UK, the designation of 'Beacon Hill' was changed to 'Henton'.

these deficiencies. The variety of areas so affected inevitably reflects a high incidence of pilot-error situations, and the relationship will become evident as specific factors are considered.

However, it will be useful, firstly, to return to – and to bear in mind – the theme of 'simplicity' as a desirable feature of the ATC system, since so many of the shortcomings to be revealed arise from the inherent complexity challenged in this study. It is the case for simplification of the system that is argued here. The primary benefits would include the reduction in error occurrence, the reduction of the possible effect of error, and the allowance of time for error correction. Simplification would also make for a drastic reduction in the number of areas in which topographical, climatic or other hazards exact the most severe penalties for navigational error; and in so doing, would permit a higher level of concentration to be applied by – and perhaps be more realistically expected from – both pilot and controller during operations in such regions.

The problems associated with air movements in the vicinity of terminal areas may be used in illustration. Here, cockpit workload tends to be high, and the aircrew, placed correspondingly under pressure by the need to practise noise-abatement procedures, make use of a diversity of navigational aids, and at the same time comply with radar-vectoring instructions. In addition, changes of radio frequency may be necessary, and extraneous information – such as 'runway visual range' (Chapter 1, p. 24) – must be assimilated.

The origins and evolution of these problems can be traced back to a somewhat grotesque anomaly; simply, the fact that few – if any – of the world's airports have ever been sited with reference to *aviation* route structure and ATC requirements. Indeed, the geographical position of airports, since the dawn of commercial flying, has been dictated by factors which are largely without relevance to their purpose. Airport locations, therefore, have been chosen on the basis of the availability of 'suitable' land, or in deference to environmental needs, or – in the case of further airport development or extension – because an aerodrome already existed on that site.

Control and operational flying techniques, as a result, can be severely handicapped by the difficulties attending any of the following:

(a) The interface problem between adjacent aerodromes.
(b) The navigational limitation of point source aids – i.e. radio beacons, markers, etc. – in terms of both route and profile navigation.

(c) The difficulty of siting such aids in the optimum locations.
(d) The long-term scale in siting such aids, when a change in route
     structure, or a new route structure, is required.

The list lends itself to considerable extension, since the changing pattern
of air transport and its needs frequently eludes prognosis; thus, the ATC
system is always behind actual requirements and is always trying to
catch up. To amplify this illustration, however, three further features
can be listed:

(e) The relatively large amount of airspace required to ensure safe
     separation between aircraft in a fully controlled environment, and
     the frequent lack of such airspace.
(f) The number of large turns required to follow the route structure,
     and the inherent danger of overshooting the turn, with subsequent
     hazard. An equally vital element is that the variation in the point
     at which the turn is initiated, and the variable rates of turn chosen,
     occasionally result in loss of longitudinal separation between
     following aircraft.
(g) The vast variation in the performances of aircraft operating in the
     same area.

Little can now be done in respect of aerodrome siting and the proximity
of adjacent airfields. However, if the airspace available is to be utilised
more safely and effectively, then aircraft navigational ability must be
improved to allow a more rational route structure to be established. If
such an improvement of navigational ability can be achieved, more
aircraft could be accepted in the system with a decreased workload for
both pilot and controller.

Figure 14 shows the extent of the problem within the London
terminal area, and a proposed solution. It can be seen in Fig. 14(a)
that, due to aircraft following the same routes, there is direct conflict between Gatwick departures to the north, Luton departures
to the south, traffic in the lower levels in both directions on the
Amber 2 airway, and all aircraft passing through the Lambourne
holding pattern. Currently, this situation is resolved by the use of
radar vectoring, and the application of vertical separation through
the Lambourne holding pattern; but ideally, the basic requirement is
for aircraft to be able to follow the tracks normally utilised by the radar
controller, thus avoiding the high controller and cockpit workload

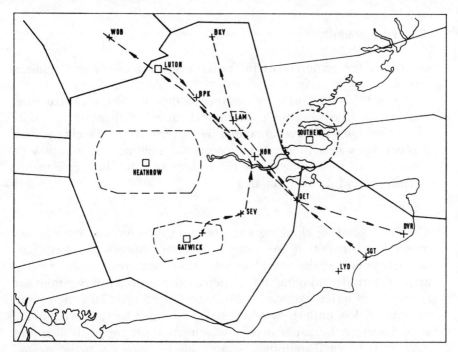

Fig. 14 a Air traffic in London terminal area, showing conflicts

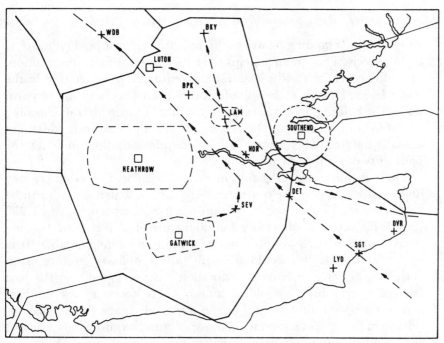

b Proposed solution showing parallel route structure

SEV	Sevenoaks	LYD	Lydd
HOR	Hornchurch	SGT	Sandgate
LAM	Lambourne	DVR	Dover
BPK	Brookmans Park	BKY	Barkway
DET	Detling	WOB	Woburn

caused by the requirement for radar vectoring on to the required tracks.

Figure 14(b) illustrates the advantages of a parallel track structure, and equally illustrates the problems which arise when a main route passes through a terminal area holding pattern. When stacking occurs, it is necessary to arrange slots through the holding area to allow en route aircraft to transit, or for the radar controller to vector transit aircraft around the holding area.

Uni-directional flow

This is essential for climbing and descending routes and offers a considerable contribution to safety. Overtaking speeds are low, particularly in view of the speed limitations imposed within most terminal areas. Controller thinking time is increased, required interventions are reduced, and pilot workload is decreased. Here again, the use of uni-directional flow patterns would ensure more time for the correction of errors; while in the case of an error which remained unperceived and/or uncorrected, visual sighting and appropriate avoiding action would normally be possible because of the low relative speeds involved.

Segregation of holding patterns from the route structure

En route. When landing delays are significant, aircraft hold en route at higher levels. This practice helps to reduce aircraft fuel consumption, simplifies diversions and makes for the reduction of congestion in the lower levels. However, the use of protected airspace and the physical size of the holding patterns results in conflict between aircraft climbing out, those in transit through the system, and those aircraft which are actually holding. This becomes a most complex situation, in which the possibility of error is greatly enhanced.

There is thus a serious requirement for the allocation of emergency holding areas clear of the route structure, within which aircraft can be held when long delays occur. Within the UK, airspace above FL 250 (25 000 ft) is controlled; as an essential feature of such control, holding areas available at 20 minutes' notice should be made available. It is appreciated that this could seriously affect military and research activities, but civil/military coordination has developed over the past few years and solutions could be found, where necessary. Certainly it is now time that all activities in the upper air space should be considered in the light of national air transportation requirements, and not as individual and autonomous operations.

Terminal Area Holding. The segregation of landing stacks from low level en route traffic, departure routes and inbound routes to adjacent aerodromes is a major requirement, but is not totally achievable in the restricted airspace of terminal areas. There are obvious restrictions in the siting of such holding areas – such as the degree of proximity to the aerodrome being served and the local environmental considerations. Once again, however, the major restriction is the reliance on point source aids, and on the ability to site such aids in the required areas.

The situation is made even more critical by the fact that the number of levels available to departing aircraft routing below landing stacks is generally limited, and the respective speeds of these aircraft can vary greatly. Because of this, complexity and congestion must arise and aircraft must be subjected to much radar vectoring. Clearly, the sooner initial climb can be achieved the sooner the problem is reduced; but early initial climbs can only be achieved (a) if the initial departure routes can be separated from the holding areas and (b) if aircraft are released from irksome and restrictive noise-abatement procedures, which, however well intended, are of very doubtful value in the context of both flight safety and the environment.

Errors in holding pattern. Two errors in the holding pattern give rise to especial disquiet. Occasionally, an aircraft entering the hold may turn in the wrong direction – i.e. take up a right-hand pattern instead of a left-hand pattern – or vice versa; and secondly – and much more frequently – a single aircraft, or a number of aircraft, may not contain the flight pattern within the prescribed area, particularly on initial entry into the holding pattern.

The hazards of the pattern-reversal situation are self-evident. In the second case, similar risks and manifold difficulties arise because ATC separation standards are based on the ability of the aircraft to remain within the prescribed area, and on the supposition that it will attempt to do so. On that assumption, en route aircraft are allowed to transit on specific routes *deemed separated* from those holding areas, without radar monitoring; other aircraft are radar directed past the holding areas in the belief that the aircraft holding will *not* stray outside the prescribed areas.

Accuracy of profile navigation

One of the basic requirements of the ATC system is the application of *vertical* separation between aircraft, on crossing or intersecting routes.

The reason for the requirement is that it is not possible to judge accurately at an early stage whether a conflict situation will or will not exist. Should a conflict situation exist, the amount of airspace/deviation from track required to resolve the situation under radar control is, in general, unacceptable in busy airspaces.

Thus, ATC clearances frequently stipulate that aircraft must reach a cleared level at a specified point. Since separation from conflicting traffic is at risk, strict adherence to the clearance is essential; alternatively, early advice to ATC of inability to comply must be given. If this indication of inability is given late, the controller may have no room for manoeuvre; levels ahead of the aircraft may be occupied, while traffic on adjacent tracks and following traffic may well preclude a turn being carried out. Controllers are emphatic that even though a pilot may be unable to comply with his clearance, there are no circumstances – when operating within controlled airspace – in which he should carry out a turn without first receiving permission to do so; for the amount of airspace used in making a turn is more likely to cause conflict with other aircraft, than would continuation on the original heading. (The origins and consequences of one such episode are described in Chapter 1, pp. 57–60.)

Ideally, this situation should never occur and is certainly considered too important a factor to be allowed to occur. Yet it must be admitted that it does occur from time to time, and for a variety of reasons. Among these are the possibilities that the ATC clearance may have been issued too late, or that the navigational aids in use may provide inadequate 'along track' information. Clearances such as 'Descend to flight level 130 to be level 10 miles before Beacon X' offer highly dubious advice; but the essential point is that any such clearance should not be accepted by the pilot unless he is certain that he can comply with the instruction. The use of the Mode C* height read-out system does reduce the danger to some extent; but not entirely, since a controller may be concentrating on another problem at the relevant time and not note the height read-out. It will be shown somewhat later in this chapter that controllers cannot monitor all aircraft under their control *all the time*; as a further complication, Mode C height read-out capability is simply not available to many controllers.

---

* The modes in secondary surveillance radar (SSR) refer to the types of aircraft transponder operation which can be selected by the air traffic controller to obtain different kinds of information about the aircraft under his control. Mode C gives the controller an automatic read-out of altitude information, telemetered from the barometric altimeter in the aircraft.

Failure of pilots to achieve cleared levels is perhaps most prevalent in the case of aircraft joining the airways system from uncontrolled airspace, or Flight Information Regions (FIR). It is appreciated that accurate navigation in such cases may be difficult, due to the lack of navigational aids within the FIR; but once again, it cannot be too strongly urged on aircrews that failure to comply with ATC instructions, for whatever reason, is dangerous.

5. Radar identification and monitoring

Aircraft are identified by radar controllers in a number of ways, but within the controlled environment where SSR is not available the most common method is the correlation of a radar echo with a position report given by the aircraft. Since other aircraft unknown to the controller may be operating in the vicinity (i.e. aircraft following the airways system, but flying visually below the controlled airspace), the controller uses additional information obtained by reference to his data display, or to 'past history' by having observed the aircraft approaching the beacon. Or, if he is still in doubt, he may request the aircraft to execute a turn, and observe the turn to confirm identification.

It is therefore obvious that accurate position reporting is essential, and that inability to offer an accurate position report must be clearly indicated to the controller. Congested R/T may preclude the making of a position report when the aircraft is actually passing over a beacon; or the aircraft may pass abeam of the reporting point. The controller must be advised accordingly.

Radar monitoring

There is a popular misconception that, once identified by a radar controller, each aircraft is under constant surveillance. Unfortunately, this is neither true nor possible. A radar controller must refer constantly to his flight data display, and to other information being supplied to him; and in addition must carry out other duties such as inter-coordination, which will require him to remove his concentration from the radar screen. The monitoring process is, however, the most important of his tasks; during this, the controller scans his radar display rather as a pilot scans his instruments when flying manually under IMC conditions. The major difference is that a pilot concentrates in great measure on a fixed number of flight instruments, and is only occasionally required to scan others. In the case of some important instruments, too, visual and/or aural warnings are incorporated to alert him to possible malfunctions.

The radar controller, in contrast, concentrates mainly on the variable number of aircraft operating under his direct radar control, and scans those other aircraft which are procedurally separated from all other traffic, and require no immediate control action, at infrequent intervals. The frequency of such scanning is entirely dependent on the workload generated by the aircraft under his direct control.

Thus, should any aircraft *not* under direct control deviate from its allocated route – for whatever reason – the controller may not immediately be aware of it.

This situation is not assisted by the practice of omitting position reports in radar environments in order to reduce the R/T load. A position report of the type 'passing Beacon X at (time), Beacon Y next . . .' can alert a controller immediately to an error in routing which he may not have noticed during his observation of the radar. In view of the indisputably heavy R/T loadings on some frequencies, it is arguable that the balance of safety is better served by the omission of position reporting when possible; but it is unfortunate that the insurance represented by these reports should be forfeited on this score. More and more responsibility is being placed on the radar controller to monitor and check the navigation of all aircraft under his control, *all the time*. If this total surveillance is to become a definite requirement, then it must be treated as an additional service to be provided, either by increased ATC staff carrying out pure monitoring duties or by a reduction of the number of activities in the radar controller's workload to allow him time for this task.

Pilots have an obvious problem in understanding exactly what service they *are* receiving. A pilot may well be advised that he is identified on radar and that he has no known conflicting traffic. However, he subsequently finds himself in visual conflict with an unknown aircraft, takes avoiding action, and complains to the radar controller that he was given no warning. The probability in such cases is that the radar controller had been actively directing a number of other aircraft some 40 or 50 miles ahead of the offended pilot, and had failed to notice the possible conflict – a situation commonly generated by the 'mix' of aircraft operating under VFR and those operating under IFR, within controlled airspace.

6. Mixing of controlled and uncontrolled aircraft under visual flight rules

The acceptance or otherwise of a mixed IFR/VFR environment is, in general, a pilot responsibility. The problems and dangers of such a mix have been discussed at length for many years, and accidents such as that described on page 149, as well as a considerable number of incidents attributable to these causes, have occurred in various parts of the world. It is clear that under these conditions the controller may be placed in an extremely serious position, since his separation may be eroded, if not destroyed, and his workload increased to unacceptable proportions.

The control system may become jeopardised in many ways; but typically, in a mixed environment, pilots operating in VMC must, on encountering conflicting traffic, obey the rules of the air and take the appropriate avoiding action. In doing so, they may be forced to disregard a controller's instruction to maintain a specific heading, and thus reduce the separation, applied by the controller, from IFR traffic on another track which may be only three miles away. The hazards of this situation may be aggravated further, since the high rates of climb and descent of modern aircraft are such that, having taken the appropriate action, and having, in consequence eroded the ATC separation, the aircraft may well become IMC and dependent on the controller to re-establish separation from the IFR traffic.

Equally serious is the approach phase, when aircraft are being radar positioned in an in-line approach sequence applying minimum separations. It is not reasonable to expect that the controller can 'increase separation' as and when required, since he may have no prior knowledge of the proximity of VMC traffic. Thus, one uncontrolled VMC aircraft may conflict with one or more of the controlled aircraft, causing a break-up of the sequence and resulting in a situation unacceptable to both pilot and controller. In the interests of a safer and more intelligent operation, therefore, it is essential that, in any 'mixed' environment, the minimum separation standards applied by ATC between controlled aircraft should be substantially increased. Such a move would allow space for visual avoiding action to be taken, and permit IFR separation to be maintained. It would also ensure that controller workload is reduced sufficiently to allow time for increased radar surveillance. However, the problem of random entry of aircraft into the controlled airspace remains one for which only one recommendation will suffice.

Positive control must be made mandatory for *all* aircraft in the vicinity of aerodromes, when public transport aircraft are operating.

7. Weather effects

Being the stuff of drama, the in-flight problems of adverse weather are well-known; less so are the effects of bad weather on the control operation, although these may be equally serious. Major difficulties brought about by these conditions include:

(a) The unduly high workload imposed on pilots and controllers.
(b) The complete cessation of landings/departures due to weather conditions falling below the required minima.
(c) A reduced rate of landings/departures due to adverse weather, or adverse runway conditions (e.g. low cloud base, snow, ice, excessive water on the runway).
(d) Sterilisation of parts of the airspace due to the need for avoidance of turbulent conditions (e.g. thunderstorms).
(e) Sterilisation of one or more cruising levels due to clear air turbulence.
(f) Sterilisation of some of the lower cruising levels due to very heavy icing conditions.

When adverse weather conditions are forecast many aircraft delay their departure; but others depart on time on the basis that, if necessary, their flight can be held over temporarily at the next staging point. Nevertheless, there are two occasions when a heavy build-up of traffic may occur; firstly, when the adverse weather conditions have not been forecast; secondly, when the forecast improvement time is premature, and aircraft have arranged their arrival time in relation to this forecast.

In such cases many aircraft are likely to be holding within the system for long periods of time. In order to plan a course of action in respect of immediate diversion, or the length of time they should hold before diverting, pilots require accurate information, either on their total delay time if landings are already in progress, or – if they are not – on their total delay time when landings commence. If landings have not yet commenced, pilots also require frequent up-dating of the forecast improvement time; but the calculation of delay times under these conditions is perhaps one of ATC's greatest headaches.

Because this service is required on relatively few occasions – and then only during certain seasons of the year – controllers are unable to gain the practice which would enable them to reach a high level of expertise. Further, the situation is subject to a number of variable quantities, for any calculation is dependent on accurate forecasting of the achievable landing rates, and the number of aircraft able to land. These variables, too, pose their multiple questions:

1. Will the adverse weather or runway conditions:
   (a) remain static?
   (b) improve, and if so, at what rate? (increasing landing rate)
   (c) deteriorate, and if so, at what rate? (decreasing landing rate)
2. How many aircraft will be able to land?
   (Aircraft landing minima vary, and if there is no improvement or only a slow improvement in conditions, then some aircraft with the higher landing minima will divert.)
3. How many aircraft will arrive, and in what order?

The final question is particularly difficult to resolve, for while many airlines at the arrival aerodromes receive notification of the departure times of their aircraft, no such information is sent to ATC. The arrival Area Control Centre is therefore not aware of the inbound aircraft until an estimate is received by the adjacent centre. The degree of warning received prior to arrival at the destination aerodrome is thus dependent on the extent of the notice given by the adjacent centre, and on the distance between the FIR boundary and the arrival aerodrome. In the UK, aircraft bound for Heathrow have experienced this problem, since the London Air Traffic Control Centre may be in contact with an aircraft approaching from the west and requesting his delay time – but be lacking in information on aircraft approaching from other directions and likely to arrive first.

The possible inaccuracy of ATC forecast delays is recognised by pilots, who tend to hope that delay times will decrease, or alternatively, that the expedition of their operation is better served by becoming airborne and holding for long periods.

In fact, this results in unnecessary and unacceptable saturation of the ATC system. The workload generated by very adverse weather conditions is already more than sufficiently arduous, due to the requirement to slot en route aircraft and aircraft bound for other aerodromes through the holding stacks; to organise and control diverting aircraft;

and to extract landing aircraft from the en route hold, and feed them in the correct order into the inner landing stacks – meanwhile continuing to supply up-dated delay times.

Air traffic control must take a firm grip on this situation, and arrange for the short haul operations and returning diversion aircraft to absorb their delay time on the ground. The ability to do this exists; but again, it is necessary to repeat that possibly the greatest shortcoming is lack of practice. Yet this type of problem exists in a number of areas of ATC, and could be much alleviated by the use of sophisticated simulation at operational units. This would provide controlled, regular, refresher training in techniques not regularly employed, and permit evaluation of the best methods of operation. Operations of this type, too, would be relatively simple to simulate since their problems are basically those of planned operations rather than of tactical direct control.

Adverse weather conditions en route

Weather returns on the controller's radar display were a common occurrence with the older type of 10 cm radar, wherein the clutter reached such proportions that radar control was often not possible, due to the controller's inability to see the aircraft. Subsequent radars, however, with sophisticated weather suppression, have almost eliminated such returns. Those which are seen do not, in any case, necessarily indicate areas of turbulence, but more frequently show areas of high water content and large droplets – these areas having a high reflective quality.

The pilot's decision to avoid areas of adverse weather is based on the indication received on his weather radar, the weather forecast, and information received from preceding aircraft, either directly or relayed by the controller. However, the avoidance of weather areas invariably presents the controller with further difficulties, since, although he suffers a reduction in the airspace available, he has neither control over the route of the aircraft nor any way of immediately reducing the number of aircraft under his control. He may also find himself with an increased workload in endeavouring to monitor aircraft which have left the protection of controlled airspace and are now operating in the FIR rather than risk the hazards of the weather.

It would be helpful, under these conditions, if controllers had an indication on their radar displays of the hard core areas which pilots would wish to avoid. This would prevent aircraft being given unacceptable vectors and permit the early arrangement of traffic pattern clear of these areas. Further research in this sphere of the control operation could well be profitable.

8. The future

For years ATC suffered from the popular belief that it was simply a restrictive organisation. It has suffered, too, from the consequences of this image, since pilots have been conspicuously slow to accept that positive radar sequencing makes for a safer and more expeditious landing flow. Even today, many pilots are reluctant to accept various forms of speed limitation, although the overall benefits of smoothing the flow and reducing the hold requirement are patently obvious. It is equally frustrating that ATC has expanded rapidly through sheer necessity, and has found itself fully occupied in merely trying to catch up with the contemporary operational situation – yet as the guardian of separation, ATC's thinking and preparation should be directed to anticipating the changing pressures on increasingly limited airspace.

Only in rare cases does ATC find it necessary to plan for future operations, or for a specific programme such as the introduction in the UK of full radar cover for the trans-sonic climb/descent phase of the Concorde operation. As a major contribution towards the rectification of this situation, companies and operators intending to fly regularly within controlled air space should be required to submit their requirements to ATC. This would enable a check to be made on whether capacity is available in the system at the required times, and would also give early warning to ATC of the necessity to provide increased capacity, or to regulate the rate of flow within the capacity available. Short-haul feeder services are particularly important in the latter context, since they frequently have unusual requirements in terms of non-standard routings, and may accordingly require special provisions.

The European theatre of operations is facing the challenge which has already been experienced in the USA, since the number of private, business, and executive aircraft operating within controlled airspace is increasing at a great rate. The executive-type jets to date have caused few problems; in fact, their relatively high rates of climb and descent and their general flexibility have made for easy integration into the ATC system. One potential problem, however, is that operation in numbers in those spheres where strict regulation of the traffic flow obtains could result in delays to the scheduled operators.

Conversely, the unpressurised but otherwise fully sophisticated private/business aircraft operating in the lower levels, (FL 100 or 10 000 ft) are beginning to cause real problems. While their numbers are not as yet great, the rate of increase is high, and a high interference factor is caused by the transit of low and comparatively slow aircraft through busy terminal areas and the subsequent conflict with local climb-out/descent routes. This is most certainly a case in which segregated routes and strict navigational accuracy should become mandatory if capacity is to be increased. Because future operations will continue to apply this pressure on limited airspace, current separation methods should be re-examined and rationalised to permit safely the higher density of traffic which can be envisaged.

In many areas the rate of flow of traffic is strictly regulated due to the lack of capacity in various parts of the system. This lack of capacity is frequently associated with the more complex route cross-over points, wherein a small number of cruising flight levels are allocated to each route. Nevertheless, there is ample capacity available in these areas provided that the required equipment and organisation is available; although the limitations on the number of levels available must remain, since vertical separation must be applied at the cross-over points.

However, either by the use of speed regulation coupled with radar monitoring, or by the use of direct radar control, the *longitudinal* separation between aircraft on the same route could be drastically reduced in safety. Subject, again, to the provision of the necessary equipment and organisation, this would call for three requirements:

1. That under normal circumstances each part of the system en route can accept the traffic being offered; or alternatively, that the number of aircraft allowed is restricted to the rate acceptable in that part of the system with the lowest capacity.
2. The provision of adequate holding areas for each route, separated from all other routes; such that, should any part of the system be forced to lower its acceptance rates severely, the preceding controller would have adequate opportunity to provide vertical separation and to hold aircraft en route.
3. Adequate back-up, or stand-by equipment, to ensure safe operation in the event of a main equipment failure.

The reduction of longitudinal separation may appear to be an operation fraught with risk, but if the proposal is to be seen in perspective, it is necessary to compare the respective – and relative – levels of safety for two conditions. In one, there is the possibility of any major equipment failure during the normal air operations in busy terminal areas wherein traffic is climbing and descending in complex patterns; in the other, there is reduced – but controlled – separation between aircraft on the same track. Subject to the provisions listed above, it is fair to say that the level of safety for the latter condition is extremely high.

The limiting factor of all operations should be runway capacity and, as a basic function of this, the traffic at all airports should be subject to strict scheduling to *below* the maximum capacity. However, the close proximity of airports in some terminal areas, and the interaction of their departure/arrival routes, results in a situation in which the capacity of the terminal area control to feed and accept aircraft offered is *less* than that of the runways available. In such cases the rule holds good; strict scheduling to *below* maximum TMA (terminal control area) capacity must be instituted.

Every controller recognises, of course, that even the most efficient scheduling will not prevent adverse weather, runway blockages and equipment failures from producing traffic build-up. But if safety levels are to be maintained, ATC must become more proficient in the art of regulating the flow to suit variable situations – and airlines, operators and pilots must assist in both planning and operation.

Air Traffic Control has been forced to develop at an increasingly swift pace during the last thirty years. Much has been achieved in that time and, if the rate of development is compared with the increase in the number of aircraft operating, the increase in their overall performance, and the improved safety and service offered, it will be seen that ATC has unquestionably gained ground. Yet it must be acknowledged that much remains to be achieved, and that the system still has a long way to go before it can truly offer the standard of efficiency which all who use the airways would wish it to reach.

This chapter has outlined some of the practical difficulties faced by the air traffic controller. It has not been coy in discussing the major inadequacies of the system which he serves. For those who are not pilots, this study will make for a greater awareness of the progress which the ATC service has yet to make; and this knowledge must be a factor in the assessment of future 'pilot-error' incidents within the system. For those who *are* pilots, an equally urgent recommendation must be made. The interdependence of those who fly, and those who must accept the responsibility of guiding them through the most critical phases of their flight, is such as to require more cooperation, fewer rejections of arduously compiled ATC instructions as mere 'dumb sheets', and fewer attempts to 'beat the system'. It is impossible to beat the system since its capacity is limited. The corollary – for the pilot who ignores ATC advice, or who breaks ATC rules – is a bill which is paid, all too frequently, not by himself, but by his colleagues, and by those in their charge.

5

Airline Management–Pilot Relations
Arne Leibing

The suggestion that pilot-error verdicts can be less than objective calls for some examination of recorded management–pilot attitudes. It is not difficult to identify their roots, nor to illustrate the influence of these attitudes on aviation safety.

Civil air transport is a commercial enterprise which shares the pressures of any other business. It shares too, industry's traditional protagonists, namely managements dedicated to achieving the maximum return on high-cost equipment, and personnel – for the purposes of this chapter, pilots – equally concerned with the maintenance of professional standards, whether of safety, performance or remuneration.

In so sophisticated an industry as aviation it may be imagined that the equation is more readily solved; yet, unhappily, it must be said that the labour relations dialogue in the regulated, specialised, and international sphere of airline operations is by no means an object lesson in the art.

Over the years, the public have been witness to some surprisingly ill-tempered disputes within the industry, and invariably the press and the media have not been slow to compare the pilot's lot with that of workers in other fields. Yet fundamental differences are involved which, seen in context, do much to explain the apparent intransigence of both parties, and perhaps the sense of frustration which once moved an airline chairman to rebuke his pilots – all of them commanding and manning a great fleet of jets in transcontinental and transatlantic service – as 'spoiled children'.

The intimate relationship between many of the 'social' and safety aspects of the day-to-day operation of big transport aircraft is typical of these differences; and it is far from widely understood how critically

this relationship influences the economy of the airline and the performance and working conditions of the pilot group. The relationship also
extends into another dimension, since most decisions concerning these
elements will influence safety and welfare, in addition to the ticket price
for the travelling public.

Many technological and economic considerations must affect the
making of these decisions, of course; but it will be shown here that a
completely disproportionate role in the matter of their acceptance or
rejection is played by the process of industrial bargaining.

The crew-complement controversy

Of the many such issues which have troubled relations in this industry
on a large scale, none reflects head-on conflict and organisational interest more clearly than does the crew-complement controversy – the
bitterly contested proposal that larger and more complicated aeroplanes justify the employment of three pilots in each cockpit, instead
of the traditional pilot and co-pilot team.

It must be said at once that there is no single all-encompassing
reason for requiring a crew of three on the high-performance jet
transports in use today. There are, however, a variety of important
reasons which, together, add up to the need for a third crew member.
One such reason stems from the fact that aircraft separation by 'see and
be seen' guidelines is still widely in force, and mixed VFR (flights
according to Visual Flight Rules) and IFR (flights according to Instrument Flight Rules) traffic is not uncommon in airspace wherein the
closure rate between aircraft may reach 1100 m.p.h., i.e. 500 metres per
second. This situation – brought about by economic considerations,
and in many areas by the large and steadily increasing group of general
aviation and private flyers – alone points the need for an extra pair of
trained eyes.

Yet despite the repetition of disasters in which the cause of a mid-air
collision is attributed to 'crew failure to see and avoid', the three-
pilot crew concept continues to generate political and industrial reaction. Other considerations aside, where there are two pilots only, the
possibility exists that one of them may become incapacitated by accident or illness during the flight; and it is not unknown for a pilot to
die of natural causes at the controls.* This point is made by Captain

* A Piper Navajo Chieftain aircraft crashed near Leeds, Yorkshire, in December 1974, resulting
in the death of all eight people aboard. It was suggested that the pilot had suffered an acute heart
attack.

Vernon W. Lowell of TWA. Writing in 1967* with more than 22 years of command experience behind him at that time, Captain Lowell discussed the 'fail-safe crew concept' and posed the question: 'Is it safe to operate with a one-man crew?' It will be seen that the answer, as long as the present airworthiness certification rules for aircraft are unchanged, remains dependent upon the industrial bargaining power of the operators and the pilots.

Beginnings

How did the crew-complement controversy originate in the first place and what makes it such a difficult problem to solve? The answers may seem to lie in the mists of commercial aviation history, and the relevance of events, discussions and decisions taken long ago may not be immediately apparent to the reader. But there is in fact excellent reason for probing this background, since it is the origins, conduct and development of this dispute, rather than any programme of technical evaluation, which have largely determined the present-day flight-deck complement.

United States Federal regulations, requiring a third crew member on all transport category aircraft weighing more than 80 000 lb, became effective in 1948 and remained in effect until 1964. This apparently arbitrary weight was established in order to exclude the Douglas DC 4 aircraft, but still require a third crew member on the DC 6, and subsequent heavier aircraft. The primary reason for the promulgation of this regulation was a series of tragic crashes involving the DC 6, which provoked public clamour for government action to make the skies safer.

The advent of the jet had become a matter of grave concern among pilots the world over, particularly in view of the early disasters which struck the first commercial jet, the British-built Comet. Here was a quite new technology incorporating operating requirements, a speed spectrum, and safety parameters which both pilots and airline management – as well as manufacturers and legislative air authorities – knew very little about.

Following this episode the UK CAA proposed the desirability of an addition to the requirement of Article 18(3) of the Air Navigation Order 1974, applicable to aircraft of less than 5700 kg carrying seven or more passengers. The requirement in such cases would be that not less than two pilots should be carried.

* Lowell, V. W. (1967) *Airline Safety is a Myth*, Bartholomew House.

It was unfortunate, but perhaps inevitable, that some of the parties involved appeared prepared to consider the new aircraft as 'just another aircraft'; but there were other views, particularly among pilots, and the 1958 annual conference of IFALPA, in Bogota, passed a resolution requiring a three-pilot crew on all jet transport aircraft in the future. Thus, in respect of this issue, the way was paved for the years of turbulent management–pilot relations which followed.

As the prime leader in aviation and aircraft manufacturing, the USA was setting the goal and influencing the entire aviation community – and since American pilots represented, and still represent, the largest national membership in IFALPA, it was no wonder that pilots the world over followed their lead. The USA was also influencing ICAO standardisation work in all areas of aviation, including airworthiness and operations, and it was clear that what happened in the USA would, in the course of time, be followed up elsewhere.

One of the main elements of the dispute arose from the fact that the requirement for a third crew member – traditionally a flight engineer – specified the borderline of 80 000 lb maximum take-off weight. All jets at that time were over 80 000 lb and, in accordance with the Bogota resolution the pilots demanded a minimum of three pilots on such aircraft. But in fact, this meant a *four*-man crew, a proposal which the airlines firmly rejected (and not only in the USA) simply because the flight engineer, the technician on board, had been a part of the original certificating process and requirement.

... The overriding public interest ...

Aviation at that time had not by any means attained the industrial systemisation of today, whereby 'maintenance production' and 'technical production planning' has displaced the former technical crew member, who often 'fixed' troubles along the line. The refusal of the airlines to countenance a four-man crew, and the insistence of the pilots on a third pilot in the cockpit, implied the possible sacrifice of the engineer; a possibility which generated much ill-feeling as it grew less hypothetical.

The pilots were accused of featherbedding jobs for their members and throwing their technical colleagues on the flight deck out of the cockpit window; and many pilots firmly believe that this discord was skilfully utilised at the bargaining table by the management side of the industry. Certainly it led to difficulties almost everywhere when contracts were being negotiated, while public attitudes towards the pilots – as demonstrated in the press and mass media – were consistently hostile.

Those pilot groups at the Bogota Conference who had shown some reluctance to support the new policy gave in under the pressure, and the three-man crew concept was adopted, i.e. two pilots and the flight engineer as on previous propeller aircraft; but on many occasions emotions grew hot on both sides, and pilots and flight engineers who for long had shared the particular friendship that only aviation can create became enemies more or less overnight.

In the USA the issue avalanched to such an extent that President Kennedy, on 21 February 1961, established a special commission, known (after its chairman) as the Feinsinger Commission. This body was set up 'to consider differences that have arisen regarding the performance of the flight engineer's function, the job security of employees performing such function, and related representation rights of the unions'. The President charged the Commission with the dual assignment 'of making a report to the President, including its findings and recommendations . . . and assisting in achieving an amicable settlement and agreement' with respect to the stated issues. The Commission submitted an initial report on 24 May 1961, and the statement which followed its receipt should be studied closely: 'It is my firm belief', said President Kennedy, 'that it is the obligation of the carriers and the unions to negotiate a final settlement of their differences on the basis of these recommendations. *They must in the light of overriding public interest resolve their differences at the bargaining table* . . . The public deserves, expects and demands that such settlement be reached.'

It is impossible to demonstrate more clearly the completely irrelevant nature of the arena chosen for the discussion of a fundamental air safety issue – which is still unresolved – for the ill-considered precedent set at that time was to have far-reaching effects. To promote 'economical, yet safe, air transportation', the Commission recommended that the turbo-jet flight deck crew be composed of three men. For those carriers which at the time operated with a *four*-man crew – three pilots and a flight engineer (a consequence of the Bogota three-pilot resolution) – the Commission stated that the reduction in crew size should be effected 'with reasonably adequate protection for the job equities of those employees who may be adversely affected'. The Commission further recommended that the flight engineers in service at the time should have bidding priority for the third seat, partly because the seat had historically been occupied by the flight engineer. In order to qualify, however, for the third seat on a three-man turbojet crew, the flight engineer would be required to take certain pilot training. In addition, all new hires were to be pilots. Flight engineers who were to serve on a three-man turbojet crew were to receive training which would enable them to act as pilots able to provide appropriate assistance to the pilot-in-command. This meant obtaining a commercial pilot's certificate and instrument rating, and in addition, specific type training as laid down by the FAA. The training would be provided at the carrier's expense but in the free time of the employee.

Despite its attempt at compromise, it can be said that the Feinsinger Commission did an outstanding job in at least creating a political platform for future work towards a solution; but its major impact on aviation history lies in its confirmation – however unintentional – that an important public safety problem could be used (and it continues to be used) as a counter in industrial negotiation.

The false orientation

It is similarly impossible to escape the conclusion that neither the industry nor the authorities had the courage, foresight and simple common sense to attack the problem as it should have been approached: namely, as a matter for objective scrutiny from the point of view of operational safety. In this light, two sentences – one from the President's statement upon receiving the initial report and one from the conclusion of the report – are of special significance. The President said: '(The parties) must in the light of *overriding public interest* resolve their differences at the bargaining table,' and (from the conclusion of the report) 'A solution in practice is what the Commission proposed and *what the public has the right to expect...*'

These sentences are fateful for aviation in that they offer evidence that the development of one of the industry's most crucial operational safety issues stemmed from a false orientation which has determined its progress in the wrong direction for successive decades. In the President's view the 'overriding public interest' was to avert the possibility of further strikes which annoyed the travelling public; and to do this by pressing the parties involved back to the bargaining table for an industrial settlement was, of course, merely a political act.

Neither the public, nor the President can be blamed for not understanding that the entire issue was one of 'overriding public *operational and safety interest*', and not just a simple labour relations dispute causing public inconvenience. It is clear, however, that at that time neither the

President nor the Feinsinger Committee (nor the airline industry itself, for that matter) was aware of the tremendous implications of the jet breakthrough. In particular, ignorance reigned on the sadly outmoded state of the air traffic services and the existing air traffic rules, both in the USA and worldwide. (Unfortunately it is common knowledge that in many parts of the world – and this applies to many airfields with high density traffic – the deficiencies are still evident.)*

Similarly, it must be said of the Commission's statement: 'A solution in practice is what the Commission proposed and what the public has the right to expect...', that unfortunately the 'solution' was not a solution in practice at all. It merely pushed fundamental operational and safety aspects under the rug in favour of a settlement which for the time being appeared to solve the problem for a strike-ridden industry. This pattern of short-term solutions to labour disputes is not unusual in industry in general. The negative effects of these policies, however, will make themselves particularly felt in such an industry as air transport which, with its specific character and environmental factors, requires an unusual degree of regulation with regard to operational and safety issues. Some major aspects of these 'negative effects' therefore deserve further study.

Policies and consequences

The three-man crew concept – in some airlines three pilots, in some two pilots and a flight engineer – was adopted and secured by contract bargaining all over the world for the 4-engine jet fleets, and, in the case of some carriers, for the twin-jet Caravelle. The Caravelle had a maximum gross weight of 110 000 lb, so that the USA's 80 000 lb rule became the deciding factor. In many other countries, however, carriers were not altogether happy to follow this rule and a series of labour disputes resulted.

Not only management–pilot relations suffered from these episodes, however, since divisions among the pilots themselves occasioned irritation and conflict. Some pilot groups like the Swiss and the Finns had opposed the three-man concept from the beginning and, in spite of the Bogota resolution, sided with the management view that two pilots only should crew the Caravelle. The repercussions were to be seen in prolonged disputes at the IFALPA annual conference for several years.

Yet one may speculate on the result if the Feinsinger Commission had been given a different brief – if, in fact, the President of the USA

* See p. 150.

had charged the Commission to make a full and impartial investigation of the problem with due regard to the state of the art at the time, the capability of man, and the operational requirements of the aviation environment.

On the basis of personal study of the subject in depth, and of intensive discussions with leading manufacturers and the legislative bodies of governmental air authorities, this author, for one, is convinced that the present twin-jet fleet of BAC 1-11, DC 9, B 737 and F 28 aircraft would most likely have been given airworthiness certificates *requiring a three-pilot crew*. Yet what kind of considerations have been brought to bear on the manufacturer?

In 1967 I wrote an article in *Aeroplane* entitled 'The blunder rate – some reflections on crew complement', and in that article made the following observation:

> Being invited to fly one of the new twin jets in February this year, I discussed crew-complement policy with the manufacturer at length... It appeared that competition made it impossible to sell a three-pilot cockpit to the operators. The first requirement on the operators' list seems to be a two-pilot cockpit before there is any possibility of further business talks. What can a manufacturer do in the light of such decisions, with 10 000 to 20 000 workers depending on the operators' orders? Obviously he makes the cockpit as the operator wants it or he goes out of business.

That the competition between manufacturers was a deciding factor in two- versus three-pilot cockpit design was further acknowledged in the Douglas Aircraft Company's concept for the DC 9 flight deck – of which it was said that 'the competitive position' dictated the two-pilot design.

Thus nullified, the earlier design philosophy receded even farther into the might-have-been when Boeing likewise joined the competitive parade with its 113-passenger twin-jet B 737.

It is obvious that the 80 000 lb rule as a deciding line between two or three pilots on the crew was wrong from the very beginning, and that crew workload during the advent of the new generation of twin-jets provided a much more rational criterion. It was to prove to be a very controversial criterion, however, and one very elusive of precise measurement. Some of the inherent difficulties of workload assessment and their implications for the pilot – and hence for the safety of the aircraft – will be examined further; but here it is necessary to make the point that, inevitably, completely different approaches to the problem on the part of management, pilots and civil air authorities have resulted in a failure to provide the aircraft manufacturers with clear direction in the matter of design requirements. This lack of direction has provided a strong economic motivation to the carriers for an increased profit which they imagine may be produced by reducing the size of the operating crew.* Lack of easily defined standards has therefore produced an incentive to be less safe – a platform which readily lends itself to the 'pilot-error' manifestation, and to the generation of labour-relations difficulties.

The absence of clear, concise and uniform regulations creates other problems besides the desire and opportunity for economic advantage. It should be noted that the first carrier to purchase a new aircraft sets the design features which are invariably costly to change (Chapter 3, p. 96). Modifications, therefore, are seldom made to the basic machine, and thus, not only is one aircraft and one manufacturer affected; so also are subsequent competitive manufacturers of similar aircraft, all air carriers who buy them, and all aircrews who must operate these machines.

Workload, the blunder – and safety

Human behaviour is only partly predictable. A trained operator in good health and not suffering from abnormal stress, fatigue or emotional disturbances can usually be relied upon to perform certain types of task with a high order of reliability.

However, from time to time, aberrations in his performance will occur, which cannot reasonably be attributed to negligence or other obvious causes. When such aberrations do occur in situations leading to an accident they are usually referred to as 'error of judgement' or 'pilot error' and the individual is considered to be culpable. It has been suggested that this interpretation of the nature of human behaviour is both inadequate and incorrect; that certain aberrations of behaviour can and do occur for which the individual should not be held responsible, and which cannot be attributed to any avoidable cause. It is clear that, if one admits such a concept of 'blunders', there is no method by which the designer of a complex system can protect himself from their occurrence in those functions undertaken by a single, unsupported human being.†

---

* But it is relevant to note the contribution which high pilot salaries make to this motivation (Editor's comment).

† See relevant text on the designer's approach, p. 107.

Human errors such as blunders frequently occur without leading to any serious consequences. What needs to be considered is whether modern developments in aviation are such as to increase the risk of incidents or accidents on their account. The ultimate limit of safety in such a case appears to be dependent on the 'blunder rate' of the individual human beings operating the system, and this, in turn, would mean – if we accept the suggestion that there is no way of either preventing or limiting the magnitude of those errors which we can classify as blunders – that we must eliminate all unmonitored human operator functions, or be prepared to accept an accident rate dependent on some function of the blunder rate.

This factor has an interesting bearing on the human workload in the cockpit and on the arguments over the two- versus three-pilots crew on the flight deck. The total task of flying the aircraft in terms of the work or functions which the crew members must perform is termed 'crew workload'; and, as in the design of the aircraft structure, there must be a factor of safety applied to the ability of the crew to accomplish the required functions. If, under normal conditions, the workload does not overtax the time or the capability of the available crew members, then a factor of safety exists to accommodate the unusual, or infrequently expected conditions.

Since crew workload is such a major consideration in assessing operational safety, it is necessary to give a great deal of thought to the methods by which it may be measured. A thorough search of the available literature indicates no standard for measuring crew workload as such, that is generally recognised within the aviation industry.

It is furthermore found that pilots do find it difficult to describe in words the workload required in flying and operating an aircraft. It is most probable that there are so many and such varied conditions under which an aircraft operates that to enumerate the functions required for any particular condition would require unusual mental gymnastics. It is also suspected that many of the functions performed in flying an aircraft are accomplished by the pilots so automatically that they are difficult to call to mind. This appears to indicate the need for the assistance of highly trained psychologists and physiologists, as well as methods engineers and pilots, if any headway is going to be made into the heart of this problem.*

Aerospace Industries Association of America and Air Transport

---

* But the 'highly trained psychologists and physiologists' have their own problems: see, in particular, Chapter 2, p. 87.

Association of America, both of Washington, DC, in their report to FAA concerning two-pilot operation of BAC 1-11, DC 9 and Boeing 737 aircraft, analysed a given flight – Washington to New York – by detailing all the functions and actions necessary to operate the aircraft over that route. By means of a computer program, they calculated the amount of eye-motion in terms of degrees of movement, and the amount of hand-motion in terms of inches of travel, for every required operation, by each crew member during the flight. Comparisons were made between the 737, 727 and Viscount aircraft. The number and complexity of the operating procedures were taken as one indication of the crew workload, and it was further expressed that the minimum eye-motion should indicate that the crew members were able to perform the required task. It may be noted, however, that the last point, according to aviation psychologists, is very debatable.

The American Air Line Pilots Association (ALPA), in their survey into the problem, seemingly did not agree with all the conclusions drawn from the Aerospace/ATA study.

ALPA, in their study of the twin-jet operation, worked on the basis that the 737 aircraft was designed to operate on 30-minute average segments. Thus, 30 minutes becomes the limit, or objective. But, in illustrating their point, ALPA used the CV 440 aircraft, as it would take roughly twice as long as long for a CV 440 to cover the same distance. For the same segment with a CV 440, therefore, one hour becomes the limit or objective. Since the number of routine functions or duties might be, for the purposes of the illustration, exactly the same, the time available to perform these functions is only half as great with the higher-speed aircraft. With a total of 60 man-minutes of work available, according to the ALPA study, all but 4 minutes 14 seconds would be absorbed by routine duties, leaving this amount of time to deal with other contingencies.

Both the Aerospace/ATA and the ALPA studies agree that 'if under normal conditions the workload does not tax the time or the capability of the available crew members, then a factor of safety exists to accommodate unusual or infrequently expected conditions'. Based on this reasoning, a high-density 30-minutes leg operating a twin-jet transport with a two-pilot crew would have a safety factor of about 7 per cent; the CV 440 on the same flight segment would have a safety factor of about 53 per cent; and if the available man-minutes on the twin-jet were increased from 60 to 90 *by adding a fully operational crew member* it would be about 37 per cent.

These studies had come about due to the disagreement over the 737 crew complement. Those operators who bought the plane, as well as Boeing, the manufacturer, considered themselves victims of the 'bargains' struck between the pilots and the airlines, and discriminated against in comparison with the BAC 1-11 and DC 9 fleets which already operated with two-pilot crews. Mutual unrest between the parties became such that, finally, a special review board was set up to deal with the matter.

The final report and recommendations from this special review board was issued in February 1969. It stated: 'We have come to the conclusion that the three-man crew should be continued on all flights of the B 737 airplane *for the life of the current basic agreement between the parties*, which has approximately another year to run.' Further, '... that in the coming 12 months there will be some flights on which the third crew member can make a significant contribution to the safety of the flight. Obviously, the great difficulty is that the need cannot always be accurately foretold in advance, and if the third crew member is not there when he is really needed, there is no way to make his presence retroactive.'

The ALPA statement that this decision climaxed a longstanding and intensive dispute in which the pilots' stand on safety had been heard and respected despite different positions by Boeing, the Company and the FAA, appears to overlook the fact that what was 'won' was in reality yet another industrial victory. It is impossible to claim a victory for safety since the competing aircraft types BAC 1-11 and DC 9 continued to be operated with two-man crews and with the industrial approval by ALPA of the contract.

The pilots

There is an old truism that it takes two to make a tango, as well as a quarrel. In view of the above discussion, therefore, it is reasonable to take a further look at the pilots' contribution to this protracted dispute.

It has been remarked that from the start a number of the smaller pilot associations had been unhappy about the adoption of the Bogota resolution in 1958. At that time they had not yet been confronted with the problem to the same degree as their US colleagues from ALPA who proposed the resolution, and had no inkling of the cold wind which was to come. Unfortunately the initial tactical victory of that resolution was never followed up by any planned strategy, and it is obvious that the proponents themselves were similarly unaware of the long-term possibilities. It must be considered a great strategic mistake that at the time of the Feinsinger Commission the pilot group failed to realise the negative implications of the report. In consequence, the fundamental operational and safety aspects of the whole problem were not established until it was too late – another factor which was to influence both concepts of safety and the labour relations scene for years to come.

Here there is a parallel with the unfortunate direction taken by the Feinsinger Commission – for if the pilots, too, had come to this realisation before the twin-jets were off the drawing boards, there is reason to believe that present cockpit design on most jet transports entering service after 1963–64 would have been based on the three-pilot concept.

However, when the leading pilot group in the leading manufacturing country so utterly failed to follow the policy which they had created and caused to be adopted internationally among pilots, it is no wonder that this body encountered great difficulties in finding its own track to steer. The author became personally deeply involved in the problem at the IFALPA annual conference in Rio in 1965, where (as chairman of Sub-Committee B) he came to realise how devastatingly fellow pilots had failed to get a real grip on the problem. At that conference the Committee thought that the proper course was for the Bogota crew-complement policy to stay in the manual, and be subjected to a further period in which to prove itself in practice; if it failed, then a new policy would have to be written. Clearly, it was pointed out, the whole question of crew complement was a dynamic one in the context of a rapidly moving industry.

The comment of one delegate at that time should be noted. He felt it was unfortunate that the organisation's airworthiness committee or member associations' aircraft evaluation committees had not given further warning of the impending introduction of two-pilot crews for the jet transport aircraft ... He hoped too, that 'what was essentially an operational problem would not be allowed to be taken up as an industrial problem as,' he thought, 'had tended to be the case in the existing and previous crew complement controversies ...'

This light had certainly delayed its dawning. Some seven years after the adoption of the three-man crew-complement policy and four years after the Feinsinger Commission's report, findings and recommendations, the international airline pilots themselves had not realised that to get what they wanted and believed to be the proper course for the industry and the travelling public, they must produce evidence and attack the right sources; namely, the rulemaking body of government, and not their labour relations counterpart, the airline management.

Conflicts, declarations, and one 'solution'

The fragmentation of opinion evident at that conference was a frustrating, but in the event, wholly salutary experience, since it revealed the need for a united and *clearly expressed* stand on the issue.

It inspired the author to submit a comprehensive study* to IFALPA, and was followed by many others from similarly concerned colleagues; and it was especially heartening to find in the course of preparing these studies that much material was available from many different and prominent scientific sources which provided substantial support for the pilots' basic policy on the matter.

One year later, at Auckland, however (the second time when chairing Sub-Committee B on the subject of crew complement) the difficulties were manifestly far greater than those of 1965. In the twelve months between the two conferences, several associations had been forced into accepting contracts with only two pilots on the twin-fleet entering service. The platform from which they had fought the issue was still weak, and the air safety authorities had discreetly ducked the matter. Effectively, the heated arguments, slamming of doors, and the whole paraphernalia of bargaining, had achieved nothing except to stir up discord between pilots and managements. It is not surprising that there was more labour unrest to follow.

In the chairman's report it was pointed out that 'there is a growing realisation within the ranks of IFALPA that the crew complement policy is a safety factor of the first priority; but also that its implementation in the future, as in the past, must be fought with such industrial means as are at the Member Organisations' disposal'.

Eleven years after the decision taken at Bogota, and after a succession of hard-fought annual conferences, the pilots' final policy in favour of the three-pilot crew was clarified at Amsterdam in 1969;† but that this apparent unanimity is vulnerable and that an important issue of aviation safety still remains a matter for industrial negotiation, is

---

* *Front Office Navigation at High Mach Numbers:* Arne Leibing, 1965.
† A resolution was passed unanimously for the crew complement of subsonic jet aircraft of any size. The resolution stated that the crew should be two pilots and a third crew member who should preferably also be a qualified pilot. The alternative solution was that he should be a flight engineer with extra operational qualifications in addition to his engineering experience.

witnessed in the British newspaper report quoted below:

> Pay increases of up to £70 a week for more than 200 pilots of British Airways were announced yesterday by the airline and the pilots' union, the British Airline Pilots Association. British Airways describes the deal ... as 'a pay and efficiency agreement ...' and claims that it will save them money because the pilots have agreed to waive certain structures ... and to having a two-pilot crew to fly its Boeing 707's and VC 10 jets, instead of three ... The new deal ends seven months of hard bargaining between British Airways and the pilot's union ...'
>
> *Daily Telegraph*, 1 November 1974

At the time of writing, clear, concise and uniform regulations are still lacking in the ICAO's Annex 6 – Aircraft Operations – and in Annex 8 – Airworthiness. Until such rules are laid down and enforced, the crew-complement factor – and its associated margin of safety – will continue to be offered for sale.

Flight and duty time: How long for how much?

Another item has had as wide an impact as crew complement on the management–pilot relationship, namely the issue of flight and duty time – a problem as old as commercial aviation itself. Attachment A to Annex 6 – Flight Operations – to the ICAO Convention offers guidelines, covering approximately two pages, on the regulation of flight and duty time for flight crews on a national basis. The guidelines are fairly simple and straightforward, and if the rulemakers had possessed the necessary political courage, these could easily have been made into a usable international standard. Far more difficult technical and operational problems have been solved by ICAO, and there is no reason why they should not have solved this one. But the few pages in the Annex remain as evidence of human incapacity to come to grips with a problem – simply because of the economic considerations which complicated the issue.

ICAO Circular 52-AN/47/4 contains a summary of how the member states have dealt with the subject, and mentions that certain limitations are obviously needed, but, as other contributors to this work have pointed out,* there are no generally agreed methods which can be

---

*See Chapter 1, pp. 27–33 and Chapter 2, p. 86.

used to measure the level of fatigue to which airline pilots are subject. The problem of regulating flight and duty time, therefore, is left to the member states.* Forty-one administrations have issued rules, and seventeen administrations have let the airlines construct these, and then approved them; but the majority have left the field wide open for the parties to negotiate and settle – once again – at the bargaining table. Pilots themselves have found it necessary to reconcile many different views, and as a result, the matter has always been a thorny issue during contract disputes.

It is fairly normal today for pilots to complete routine working periods of 14–16 hours, that is, fully twice as long as the average office day. By adding another pilot or crew member to the crew and providing at least a comfortable seat for rest, duty periods of up to between 20 and 30 hours are considered to be in order. In fact, a tremendous number of variable combinations of flight and duty time limitations exist today in the industry and all of them mirror the particular route structure operated by the airline concerned.

However, at contract negotiations over the years it has been interesting to note that as the going headed towards a climax, the flight and duty time term of the contract more often than not became the critical issue. The declared readiness of the pilots to go to the lengths of strike action on this issue frequently encouraged the operator to 'buy' exemption from any adjustment of existing flight time by offering some other improvement in the pilots' conditions of service; many salary increases and improved insurance schemes owe their origin to this gambit.

Yet strong as the pilot group may sometimes appear to be – and particularly during contract negotiations when the headlines are splashed across the front pages – and strong as they undoubtedly have proved to be in many respects, it must nevertheless be said they they have never succeeded in achieving a proper jurisdiction over this subject on an international scale. The greatest effort to rid aviation's labour relations scene of this blight was made at the International Labour Organisation's Ad Hoc Meeting in Geneva in 1960, which discussed the possibility of establishing an international standard of duty time for flight crews. The Conference and the pilots failed, simply because a majority composed of government and airline representatives refused to accept that there existed any need for such a standard, and

* See Chapter 1, p. 32 fn., p. 33.

clearly declared so. Admittedly, the problem is a complicated one due to the particular problems involved, such as changes in time-zones which give rise to metabolic disturbances. There are the hours of darkness, weather, number of landings, temperature changes, low content of oxygen, and problems of dehydration, etc. But nevertheless it seems appropriate that by now – some 30 years after the adoption of the Chicago Convention – a fairly usable standard should have been established for aircrew working hours. When one considers the fact that the same governments and the International Labour Organisation have found it necessary to establish, and have succeeded in establishing, legislation and international maxima on duty hours for factory workers and hotel chambermaids, the Geneva report's treatment of this subject in respect of flight crews is discouraging.

Clearly manifest in this document is a monumental failure on the part of governments and airlines to create a firm and equal standard for one of the most controversial and disputed elements of flight safety; and it is a personal – although by no means isolated – view that the political lip service contained in that report makes it a classic of its type. The relationship of this issue to the pilot-error syndrome should be clearly stated here. At the Annual Conference of IFALPA in Rio, in 1965, the following recommendation was adopted: 'It is recommended that IFALPA representatives at ICAO and other bodies be requested to insist that the expression "Pilot Error" should not be used out of context when "Pilot Tolerance" is actually meant ...' If the pilot's 'tolerance' is exceeded and an accident results, it is pertinent to question the stresses which the system has imposed on its human operator. It is impossible to know how many accidents over the years owe their origins to the fatigue state of the crew – but that this cause may have played a major part in many of them it is reasonable to assume.

There is, therefore, real cause for concern in the realisation that aviation appears to be the only form of public transportation in the world wherein the matter of determining safe operation in respect of crew complement, maximum working hours and minimum rest periods, is left to the strength with which employers and employees respectively can 'defend' their economic interests. How then has this problem been dealt with at the bargaining table? Undoubtedly, the greatest success has been achieved by US pilots.

Based on a flight-time scale of 85 hours per month/1000 hours per year, they have created a system over the years whereby the economics are coupled to the productivity of the aircraft type flown; at the same time the arrangement directly limits the maximum duty time which can be performed. The degree of stress represented by these hours has been noted elsewhere in this book. It may be repeated, however, that 70 hours of flight deck duty on a jet-liner are more than the equivalent of the usual 40-hour-a-week job in terms of fatigue, and particularly of responsibility. It was more than 30 years ago that the US airlines agreed to limit monthly flight time to 85 hours. That maximum remained in effect up to the sixth year of jet service, when the airlines admitted that the jets had created new fatigue problems not anticipated earlier. The reason lay in the obvious differences in speed between the jet-liner and the piston-engined aircraft. A pilot flying a piston-engined machine could do five transcontinental or transatlantic round trips a month to get his 85 hours; but on the faster jets, 85 hours would add up to 10 flights a month. After prolonged disputes, other pilot groups, notably the Australian and British, succeeded in obtaining similar contracts, but, in the wider international context, the situation is far from satisfactory.

The Scandinavian experience reveals the mutual depth of feeling on this problem. In 1966, for instance, it became the final stumbling block, and a strike threat was translated into fact. The Danish and Norwegian governments intervened, forced their own pilots back to work by law and made pleas to the Swedish government. In subsequent discussions between that government and the Swedish pilots, the latter agreed to go back to work, and to accept arbitration – with the condition that an investigation committee should be formed – in order to bring about legislation for flight and duty times. The Swedish government was shocked, to say the least, to find that this subject – contained (however inadequately) it will be remembered, in the ICAO agreement – had been used over the years as a ploy in pilot–management negotiations. The cheers from the pilot group were loud and clear when this condition was accepted by the government; but subsequent events showed that the pilots cheered too soon.

The committee was set up in December 1966 and, at the urgent request of both Denmark and Norway, became a Scandinavian Committee under Swedish chairmanship and in accordance with Swedish procedures. The author was selected as representative for the Swedish pilots, and worked on the committee for three years, until it became necessary to resign on entering the Flight Safety Department of the Swedish Board of Civil Aviation.

The first meeting of the committee was dramatic, since the representatives of the Danish and Norwegian governments, the scheduled
side of the airline industry as well as the non-scheduled side – with
the exception of the Swedish non-scheduled representative – roundly
declared that there was no need for such legislation and that they were
to oppose its creation. It is therefore not surprising that a further six
years of work were required before the committee found itself able to
produce a *Draft of Law Regarding Flight and Duty Time for Personnel*
(Flight Duty Time Act). Of the 15 members of the Committee, 8 were
in favour of the law and 7 against. The latter were at least – and at
last – in agreement with the majority on the need for regulating rules,
but claimed that these should be of an administrative character only
– a near impasse which clearly reflects the problem. It is not without
relevance that the reference list of the report contains no less than 99
internationally published papers on various aspects of the subject.
Material on this scale is surely convincing evidence of the need to
establish an *international* standard in the field; and of the need, similarly,
to remove this issue, too, from the bargaining table for the benefit of all
concerned. No apology is deemed to be necessary for repeating that
not least among these is the travelling public.

Accidents: Management–pilot reactions

The one area of management–pilot relations in which emotions tend
to grow harsh and bitter, and where some real backlash has been
experienced, is in the immediate aftermath of an accident, where the
airline management, for reasons unknown, or to be conjectured, have
been unduly hasty in focusing public attention on the pilot. In many
cases, the latter's death affords him no protection from this kind of
attack. It is difficult to understand why these things happen, and why
managements have so often proved themselves reluctant to stand
behind their own captains and crews until the case has been reasonably
investigated. But one root may lie in the rather naive belief, shared
by many managements, that the public demands, and must be given,
a quick answer.

This is sadly to underrate the public intelligence. The public by now
has accrued its share of cynicism, and has been exposed for a considerable number of years not only to the triumphs, but also to the follies and
difficulties of the aviation industry. It will become progressively less

likely to accept the pilot-error verdict without substantial evidence, and in the face of a growing capacity to evaluate the contribution made by other factors – such as poor maintenance, or inadequate navigational aids – which may become public knowledge.

The pilot associations, for their part, have been criticised (and with some justification) for their often one-sided and almost totally subjective approach to crash investigations. It is true that their representatives on accident-investigation teams often appear to be solely concerned with efforts to defend the pilots from charges of error; although it is not difficult to find some justification for this attitude. An erring pilot is all too frequently the victim of sins of omission and/or commission perpetrated by somebody outside the cockpit.* Fellow authors have made the point that very few crashes have a single cause; and that most are the culmination of circumstances and mistakes which cannot always be laid in the lap of the pilot who makes the final error. The pilots' organisations who invariably try to defend their members therefore ask, in effect, *why* an accident occurred, and not merely *how* it occurred;† and they do so because the history of crash investigation is replete with accounts of pilots who were blamed for accidents, while no effort was made to examine the underlying reasons for the disaster. Unfortunately, events have shown that this knowledge – widespread as it is throughout the aviation world – has done little to prevent the emergence of 'authoritative' although entirely unfounded post-accident pronouncements,‡ the effect of which is to imply human (i.e. pilot) error, rather than technical or administrative malfunction.

The crash of an SAS Caravelle during its approach to Ankara airport in January 1960, generated this familiar immediate reaction – that 'there was no reason to believe that any technical failure in the aircraft had been a factor in the crash'. While serving the purpose of reassuring the public on the reliability of this aircraft, the statement clearly threw the responsibility for the accident on the pilot concerned; but in fact, the aeroplane at that time had not achieved the technical excellence of its later series, and the need for various modifications was being voiced by its pilots. A specific deficiency which had been encountered arose from the finding of metal chips in the servodynes, by which the steering impulses are transformed into movements of the

* Cf. Dr Walter Tye's statement, pp. 63–4, and quotation from Dr R. R. Shaw, p. 137.
† Cf. Dr Allnutt on this theme, p. 66.
‡ See examples of 'the instant judgement', pp. 46, 50.

rudder surfaces. There were certain 'fail-safe' devices in the steering system, but if the servodyne became blocked it could result in blocking of the control movement, and there was no separate wire system to which the pilot could resort. The pilots were naturally ill at ease when the servodynes of the crashed aircraft were found to be suspect in this way, and it became necessary for the airline to overhaul all new deliveries of these units as a precautionary measure before installing them for service.

Fourteen months later the Turkish authorities made the report public. The conclusions were remarkably short and came as no surprise to the pilots. 'A number of factors – known and unknown – had taxed the three pilots above their capacity, and a technical failure could have been a contributing factor to the accident.' It is to the credit of the airline president that he apologised to the pilot group for the statement made after the accident.

This case is not unique, since it mirrors a fairly general behaviour pattern manifested by this problem. One may say that, as pilots are human, so also is management; and that they also respond to the pressure of events. It is no excuse – but it may offer some explanation of why these things happen.

Captains arraigned

Captains Hugh H. Hicks and Stuart A. Dew of the Civil Air Transport (CAT) of Taiwan, were at the controls of the Boeing 727 jetliner which crashed near Linkou, Taiwan, on 16 February 1968, during a routine ILS approach. Of the 63 persons on board the plane, 21 were killed. Within a month of the accident disturbing news reached IFALPA Headquarters in London, reported by the Hong Kong Pilots Association and other members operating in South East Asia.

The accident investigation conducted by the Civil Aviation Authority of Taiwan assessed both pilots as criminally negligent; as a result, they were put on trial in the District Court in Taipei, facing possible prison sentences of up to five years.

IFALPA's inquiry into the extremely detailed information provided by the American lawyer defending the two pilots indicated that the aircraft commenced its descent before reaching the outer marker, and as a result, struck a hill 850 ft above airport level, 9 miles from Runway 10 at Taipei airport. At a glance, it looked like another case of pilot error; but careful analysis of the flight recorder indicated precise course and speedkeeping, and an accurate height maintenance prior to last descent. Voice recorder confirmed a routine ILS approach, and it was only realised some 5 seconds before impact that the aircraft was below the glide path. An overshoot was initiated, which, however, could not avert the disaster. The most significant point of the flight recorder trace was that the aircraft paralleled the glide path about 1300 ft below it. This pointed towards possible interference or malfunctioning of the glide-path transmission.

During the course of the next months, this was confirmed* by several pilots' reports of false reception of both glide path and outer marker in the position where the ill-fated aircraft commenced its descent. It was believed that marker reception could have been triggered by a powerful radar station nearby.

Under considerable pressure, almost two months after the accident, the Chinese authorities permitted an FAA flight check on the ILS installation, which indicated a serious out-of-tolerance condition of one of the glide path transmitters – a disclosure which was published in *Aviation Week and Space Technology*, 10 June 1968.

One may ask where lay the onus of liability: with the pilots? – or with those who undertook responsibility for the ILS?† The reaction of the airline, CAT, was to say the least, negative. The reaction of the Taiwan–Chinese authorities, however, was quite different, and the following is quoted from the editorial column of the *Taipei China Post* of 1 June 1968. It reads:

Jan Bartelski, President of the International Federation of Airline Pilots Associations, reportedly said a few days ago that its members might boycott Taipei if Captain Hugh H. Hicks and Stuart A. Dew of the Civil Air Transport (CAT) were convicted by the Taipei District Court. Both American pilots were at the controls of the ill-fated Boeing 727 jetliner which crashed near Linkou on February 16 ...

As a matter of fact, the threat was made prematurely. The District Court has decided to hold another debate on the air crash case. It was originally scheduled to announce its verdict last Monday. (But) ... even if both American pilots were convicted, they would be able to appeal to a higher court against their conviction.

We know well that some foreign pilots might be scared away if Captains Hicks and Dew were convicted. According to many legal experts here, if a pilot has committed an error which led to an air crash, the Civil Aviation

---

* But see p. 195 fn.
† Captain Leibing echoes Captain Bressey.

Law instead of the Criminal Law should be applied.* The severest possible punishment stipulated in the Civil Aviation Law is the revocation of a pilot's licence. [Recommended in the report quoted below. Ed.]

The fact is that Bartelski's threat may be considered as interference with the Republic of China's internal affairs and as an infringement on its judicial independence, the reason being that the threat was made before the conclusion of the case. The Republic of China is an independent country and has its own laws. The two American pilots have been indicted and tried under Chinese law because the accident occurred on the territory of the Republic of China. Whether the law does, or does not, conform to international practice is another question ... It is true that no airline pilots have been held criminally responsible for their personal errors which led to air accidents. †

The *Aviation Week and Space Technology* report quoted the 'impossibility of outside intervention' in a case placed 'under the jurisdiction of the Government'. This loss of support must have been bitterly felt by pilots during one of the most crucial cases in modern air transport history – a case wherein the question of pilot responsibility for the deaths of 21 people was to be put to the test.

Captains supported

In fairness, notable exceptions to the rule are on record, and it is refreshing to be able to quote cases in which the pilots deservedly received honourable and courageous support.

Captain Vernon W. Lowell's book‡ deals at length with United Airlines defence of Captain Gale C. Kehmeier who survived the crash of his Boeing 727 at Salt Lake City airport, on 11 November 1965. Responding to the Civil Aeronautics Board's official report which

---

* See Chapter 6, pp. 200–1, for clarification of 'Aviation Law'.

† Editor's note: This acrimony, the author's statements, the disclosure in *Aviation Week and Space Technology* and the conflicting report of the CAA, Republic of China (ICAO Circular 96-AN/79) again typify the subjective elements and consequent difficulties facing the researcher into pilot-error episodes. The Chinese report (undated) states (Section 1.8, Aids to Navigation): 'The equipment was in normal operating condition at the time of the accident ... A periodic flight check was carried out on 15 January 1968 and the system was found to be satisfactory. Furthermore, on 16 February 1968, within a period of two hours before and after the accident, six other aircraft had smoothly approached and landed at Taipei International Airport using the same instrument landing system.' Section 2.1 (Analysis) again refers to the six smooth landings and adds the categorical statement: 'This accident was not due to any deficiency of the Instrument Landing System.'

It is unfortunate that, despite enquiry, evidence from the FAA was not available before this book went to press.

‡ *Airline Safety is a Myth, op. cit.*, pp. 187–93.

claimed that the 'probable cause' of the accident (in which 43 people died) was 'pilot error', United's Senior Vice-President E. O. Fennell offered his own company's report which firmly rejected the CAB's reflections on his pilot's proficiency. The final paragraph of the United Airlines report (dated 8 April 1966) epitomises our own theme:

> The conclusion fairly to be drawn from Captain Kehmeier's training and operational flying record is that his proficiency to fly the Boeing 727 aircraft as a pilot in command has been adequately demonstrated to both the Federal Aviation Agency and to United Airlines. To distort this record after an unfortunate accident – to make it appear that it gives some indication that he lacked proficiency – does not serve the goal of aviation safety.

Certainly this record must include the public defence of Captain Donald J. Cook Jr by Mr Charles C. Tillinghast, board Chairman of Trans-World Airlines, against accusations from no less a person than the former Director of the FBI, J. Edgar Hoover.

Captain Cook was the TWA pilot who saved his jetliner, his crew, and his passengers from a mentally disturbed hijacker armed with a carbine, a knife and a pistol – and also incidentally, from a pack of FBI agents at the Kennedy International Airport, who made a clumsy if well-intentioned attempt to rescue him. However, the Captain was subsequently sufficiently impolite to say of the FBI: '(Their) plan was damned near a prescription for getting the entire crew killed and the plane destroyed.'

The consequence of this *lèse majesté* was a letter from the Director of the FBI to Mr Tillinghast, the airlines board chairman, 'touching, among other things', in the words of Mr Tillinghast, 'on Captain Cook's "difficulties in the Air Force" prior to his employment with TWA.

'What these difficulties – if any – may have been, we do not know. We know, however, that if Mr Hoover had access to such information as a result of his official position, he certainly had no business making use of it for personal reprisal.'

Mr Tillinghast thus showed himself to be just the kind of board chairman who deserved the kind of pilot typified by Captain Cook; calm, level-headed, and tough enough to treat attempted intimidation in the only manner worthy of human dignity.

But these cases, it must be repeated, are exceptions, and one can only wonder at the distressing frequency with which airline managements have let their pilots down. The opposite extreme was illustrated in the aftermath of the collision between the ships *Andrea Doria* and *Stockholm* off the Nantucket light vessel on 25 July 1956, for, during the legal proceedings that followed the collision, each line claimed that its own officers were entirely blameless ... and those of the other wholly to blame ...*

The contrasts emphasise the complete lack of tradition in this field of the managerial ethos. It is legitimate to ponder how differently the management/pilot relationship might have developed: how much injustice might have been avoided; and how much more positively aviation safety might have been influenced, had such a tradition been fostered among the leadership of the air transport industry. It is surely right that the perceptive and conscientious pilots and managers at the heart of international aviation today have made a beginning.

* K. C. Barnaby, *Some Ship Disasters and their Causes*, Hutchinson, London, 1968, p. 139.

6

Legal Implications
N. D. Price

Introduction

The purpose of this chapter is to examine some of the legal implications arising from the suggestion, or identification, of pilot error as a
factor in air accidents. These implications are considered in terms of
legal vulnerability and therefore have equally forceful relevance for
all those – whether designer, pilot, operator or passenger – who are
concerned with professional aviation. None of the examples which
follow is intended to levy judgement on any individual or organisation. The intention is wholly to offer some guidance on the nature
and possible application of the law in a particular situation.

It is firstly necessary to grasp the fundamentals of a legal system –
the manner in which laws are made and developed – and the legal
relationship of those persons involved in any dispute. Without such
knowledge, misconceptions arise and the point at issue is missed.

To avoid conflict with the law, a person either consciously or sub-
consciously questions his own actions, firstly in the light of his duty to
the state, and secondly in the light of his duty to those persons with
whom he has a legal relationship.

The first aspect is covered by the *criminal* law, which regulates the
overall conduct of our society so that persons who offend against it
are punished. The second aspect is covered by the *civil* law, which
provides that persons who suffer injury or damage by the wrongful
acts of another receive redress. These, of course, are separate functions;
each must therefore be regarded in a separate light. They are not,
however, divorced from each other, for the same action may be both
a crime and a civil wrong. If a pilot flies his aircraft without regard

to the safety of navigation, and someone is injured as a result, he may be punished by the state through a criminal prosecution, and have a claim for damages awarded against him in a civil action brought by the person injured. (See cases described on pp. 221, 237; also p. 224.) It must be pointed out, however, that the commission of a crime is not, per se, a civil wrong, any more than a civil wrong is a crime. A passenger cannot sue the pilot if he is in some (criminal) breach of the Air Navigation Order* which causes no injury or damage. Nor need the pilot fear a prosecution because he has failed to complete some term of a contract which is subject to civil law. Nevertheless, the two sections still have some application to each other. The fact that a pilot has committed a crime in the operation of his aircraft could be evidence that may tend to establish a civil wrong. The fact that a person was injured as a result of the pilot's negligence could well be used as evidence that the methods of the pilot were criminal.

How laws are made

In most developed countries there are two sources of law, namely the statute law and the common law. The statutes are written laws passed by the legislature: the assemblies consider the wrongs that are to be prevented, and draft written laws which are best designed to prevent their occurrence; albeit that in their endeavours to control the functions of society, these statutes are often profuse and complex.

The common law of the country, in contrast, evolves with the passage of time following the experience of the courts. As a result, common law rulings will only be found in judgements which follow legal disputes. Sometimes they are difficult to find and the lay person is frequently confused to learn that the actions of the driver of a horse and chaise in bygone years can indeed be relevant to the twentieth century operation of a Jumbo jet. Yet this apparent incongruity is clarified in the case of *Clay v. Wood* (1803), wherein the facts were as follows:

Clay's servant was riding Clay's horse on the wrong side of the road. Wood's servant was driving the chaise out of a side road and in trying to get on to the correct side of the road (on which Clay's horse was being ridden), the shaft of the chaise struck the horse, breaking the animal's thigh. Clay sued. Wood defended himself by claiming the accident would not have

* A British statute detailing procedures for the safe operation of aircraft. See pp. 238–9 for 'typical' regulations contained in the Air Navigation Order.

occurred if the horse ridden by Clay's servant had been on his proper side. Clay won the case. As the judge said, 'if a collision can be better avoided by going on the wrong side, it is not merely justifiable to do so but obligatory'.

Today this case might be decided differently, but one fundamental principle would remain unchanged, in that everyone – including pilots – is under an obligation to take whatever steps are necessary to avoid a collision: even if the danger is the result of the wrongful actions of another, and the necessary steps are non-standard and perhaps even criminal. The application of this ruling to aviation is obvious. It follows, too, that the more intimidating consequences of a pilot's error reflect his responsibility, and not necessarily his liability.

Aviation may be a highly technical and complex industry; but if the principles under dispute have been decided in earlier precedents, there is no reason why the legal results should differ.

The common law is fluid, however, and will adapt itself to changing circumstances: if the principle is not the same, the courts will not follow. It is true that on occasion the common law becomes trapped by its own decisions, and anomalies result, but it is here that the legislature comes to the rescue and, by statutes, corrects the matter. Frequently, the anomalies are anticipated and regulating statutes develop with the technology. It is therefore to the statutes that reference must be made; if there is no covering statute, the common law rules apply.

In this chapter, aspects of pilot error are considered under the two heads of the criminal law and the civil law. There will be further subdivision so that the position of the different classes of person involved can be considered.

Air law

It is perhaps necessary at this point to refer to the term 'air law', since this implies an entity separate from other laws. The law is a homogeneous thing and cannot be so separated. A civil wrong is a civil wrong no matter how it is inflicted, and references made to 'air law' usually describe applications of the general law to an aviation incident. For example, if a person is negligent and causes injury to another, then he will have committed a tort. The vital question to be answered is whether the tortfeasor (person committing the tort) was in breach of his duty of care to the injured person. However, the same legal steps are followed whether the act complained of is a surgeon's slip of the scalpel (medical law), a driver's misuse of the brakes (motoring law) or the pilot's use of the flaps (air law). It is often more convenient to seek precedents from the same field of operation but this is not to imply that there is a separate set of laws. These pages offer many examples of cases decided long before the invention of the aeroplane, and there is no reason why their principles of law should not be included in air law if that term is insisted upon.

The foregoing should be qualified in that certain statutes refer specifically to a particular field of activity. Thus, there is no need for a doctor or motorist not involved in aviation to study the Air Navigation Order, any more than a pilot need study specific parts of the British National Health Act. Similarly, the Warsaw Convention on carriage by air concerns only aviation. These special statutes are, however, only incidental to the general laws and a reader of law would be very wrong to narrow his studies.

It is, of course, true that a number of lawyers have been – and are – acclaimed as experts in air law. It is not the purpose of this chapter to belittle the title but to explain that such experts are persons who have made a study of the law as it applies to aviation and who spend a greater proportion of their professional life in dealing with the legal matters arising from flying. Further reference is made to this topic in later pages, within the section dealing with Public Enquiries and Litigation (pp. 247–50).

Civil law

Civil law must be further subdivided into the fields of contract and tort. Tort deals with the situation not governed by any agreement in which a person, the pilot, through some legally wrong action, causes injury to another, or damages his property. The contracts deal with the specific agreements made between persons, in which the agreements govern the obligations, duties and responsibilities of the parties to that agreement.

The law of torts

Liability within the law of torts can arise either as a result of what is known as strict liability, in which the plaintiff need only show that he has suffered damage as the result of the defendant's extra-hazardous activity, or offer proof that the defendant has, through some wrongful act, cause the damage complained of. By the very nature of aviation, the international pilot is in difficulties because the laws of different countries vary. With exceptions that are irrelevant to aviation, liability in the English law of torts is based on *default* whether deliberate, negligent or reckless. The other extreme is the French Civil Code, para. 1382, which states: 'A person who has control of an object is responsible for all damage caused by that object.' It will be seen, too, that even within the USA the States are not at all united on this matter of liability. Three cases are offered by way of illustration: two from the state of New York and one from West Virginia:

D.C.N.Y. 1955 Flight of an airplane at a proper altitude is lawful, but the person operating it is charged with the responsibility of preventing injury to persons and property beneath; and not to prevent such injury, whether negligent or not, renders the operator liable at law on the theory that it was his duty to prevent it if he undertook to operate the plane.
(*Hahn v. U.S. Airlines 127 F.Supp. 950*)

D.C.N.Y. 1954 Neither the law of New York nor the law of Connecticut classifies aviation as an ultra-hazardous activity, and no absolute liability is imposed on airplane owners.
(*D'Aquilla v. Pryor, 122 F.Supp. 346*)

D.C.W.Va 1953 One who flies an airplane is opposing mechanical forces to forces of gravity and is engaged in an undertaking which is fraught with gravest danger to persons and property beneath if it is not carefully operated or if the mechanism of the plane and its attachments are not in first-class condition. At Common Law the hazardous nature of this enterprise subjected the operator* of the plane to the rule of absolute liability to anyone on the ground who is injured, or whose property is damaged as a result of this operation.
(*Parcell v. United States 104 F.Supp. 110*)

The States which have strict or absolute liability attached to aviation cause little complication to the law. If the aircraft causes the damage, then the owner or pilot should pay, irrespective of his innocence – although various defences may be available to him, such as an act of God, or the intervening act of some third person outside his control. Because of the considerable complications caused by some countries' requirement for *proof* of wrongdoing or default as a pre-requisite to a claim for damages, these rules will now be examined. For the sake

---

*The American use of the word 'operator' seems to be in the context of the operator of the controls of the aircraft, i.e. the pilot, whereas the word as used in the British sense describes the owner of the aircraft who uses it to 'operate' the airline.

of simplicity, the English law is taken as the working example, but it will be found that the law of New York is very similar. There are also a number of illustrations, taken from other countries, which also fit the English law.

Definition of negligence

Aside from intentional wrongs, which are not discussed, a pilot's error is rarely legally wrongful unless he is negligent.

What, then, in the eyes of the law, is negligence? There are two classic definitions and both will be examined:

> Negligence is the infliction of damage as a result of a breach of a duty of care owed by the defendant to the plaintiff.
> (per Lord Wright in *Lochgelly Iron & Coal Co. v. M'Mullan* (1934))

> Negligence is the omission to do something which a reasonable man, guided upon those considerations which ordinarily regulate the conduct of human affairs, would do; or doing something which a prudent and reasonable man would not do.
> (per Baron Alderson in *Blyth v. Birmingham Waterworks Co.* (1856))

Ingredients of negligence

There are six ingredients: (1) a duty of care situation, i.e. a recognition by law that the careless infliction of the kind of damage in suit, on the type of person to which the plaintiff belongs, by the type of person to which the defendant belongs, is actionable; (2) the damage suffered by the plaintiff was the foreseeable result of the wrongful act of the defendant; (3) proof of carelessness, i.e. breach of the duty owed; (4) the connection between carelessness and the damage is not too remote; (5) the extent of the damage attributed to the defendant; and (6) the monetary estimate of that extent of the damage. All these ingredients are examined in some detail below: but it should be noted that ingredients (1)–(4) must first be proved before (5) and (6) become relevant.

Ingredients (1) and (2): A duty of care and the problem of foreseeability. By the very nature of aviation, a pilot owes duty of care to a great number of people. Foremost are his passengers; but his responsibility goes a great deal further. He owes a duty of care to his fellow crew members, to the owner of the aircraft he is flying, and to the shippers of cargo carried therein. He also owes a duty of care to those persons or goods not carried in the aircraft – either on the ground, or in another aircraft, or on the sea – who may be injured by his actions.

Arising from the judgement of Lord Atkins in the case of *Donoghue v. Stevenson* (1932) came the neighbour principle. In his words:

> The rule that you love your neighbour becomes in law, you must not injure your neighbour, and the lawyers' question, 'Who is my neighbour?' receives a restricted reply. You must avoid acts or omissions which you can reasonably foresee are likely to injure your neighbour. Who then in law is your neighbour? The answer seems to be those persons so closely and directly affected by my acts that I ought reasonably to have them in contemplation as being so affected when I am directing my mind to the acts or omissions which are called in question.

Bearing in mind the broad spectrum of persons who will come within the neighbour principle, a pilot must be well aware of the consequences which his carelessness might bring about. The extreme example might be that of a careless pilot who causes his fully laden Jumbo jet to collide with another fully laden Jumbo jet over the City of London. Each and every person so injured, and the owners of the property so destroyed as a result of the accident he could foresee would follow his carelessness, have a right of action against him. Such an extreme example would be a national disaster of the greatest magnitude. None the less, in theory, the wrongdoing pilot remains liable. A number of aircraft *have* crashed into built-up areas; and a number of passengers and crew members have been hurt during the sudden avoiding action necessitated by the careless navigation of another aircraft. In each case the pilot whose carelessness brought about the error is liable for the damage he has caused.

An American case demonstrates the neighbour principle:

DC Ill. 1957 'In administratix' action for wrongful death of railroad employee, killed when the canopy of a military aircraft from which the pilot was baling out, fell and struck the employee while he was eating his lunch on a railroad right of way. Evidence disclosed that the pilot was negligent in not going around storm clouds and in failing to take proper precautions to prevent losing control of aircraft after entering the overcast. An air force base was negligent in not giving the pilot warning of the thunder storm; and all of such negligence was a proximate cause of deceased's death. (*Bright v. United States*)

There are more classes of persons the pilot can reasonably expect to be involved by his errors. He can foresee that, in the event of a crash, there will be rescuers who·may become injured or die in their gallant attempts. In 1968, a Boeing 707 caught fire in the air over London and returned to land in flames. This resulted in the death of a brave young stewardess who gave her life trying to save passengers. She could have escaped without serious harm to herself, but chose instead to risk her life for others. It is very doubtful that any court would hold her responsible for her own death. If it was established in law that a breach of duty was responsible for the fire and what followed, then the person responsible for that breach would in turn be responsible for a breach of duty owed to the stewardess. The difficulty of proving such a breach of duty in these circumstances is discussed later. A rescue attempt which was no more than a hopeless suicide would be a different matter.

In *Chadwick v. The British Railways Board* (1967), Mr Chadwick went to the rescue at the scene of a railway crash which occurred due to the negligence of a railway employee. Whilst Mr Chadwick suffered no physical injury as a direct result of the crash, he was so sickened by the sight of the devastation that he suffered nervous shock. The court held the Railway Board to be responsible for Mr Chadwick's suffering.

However, in the case of *Bourhill v. Young* (1942), a motor-cyclist, by his negligence, caused a collision and was killed. The plaintiff, a pregnant fishwife, was dismounting a tram some 45 yards away from the point of collision. She could not see what happened, but heard the crash. This noise so affected her nerves that she became ill, gave birth to an idiot child, and was for a long while disabled from carrying on her trade. The court held that although the deceased might reasonably have been expected to have foreseen injury to persons in the immediate vicinity of the place of impact, he could not have been expected to foresee injury to a person so far from this spot as the appellant: the motor-cyclist was therefore not liable since he held no duty to the unfortunate woman.

The Supreme Court of Nova Scotia also refused to hold that the pilot of an aeroplane on a regular route was under any duty to avoid a ranch containing 'noise-conscious' mink. *Nova Mink Ltd. v. Trans-Canada Airlines* (1951).

The damage suffered by the plaintiff was the foreseeable result of the wrongful act of the defendant: At English law, there have been two conflicting doctrines: the *directness test*, which arose from the decision of the House of Lords in *Polemis & Furness, Withy & Co.* (Re) (1921), and the *foreseeability test*, formulated in a Privy Council judgement of *Overseas*

Tankship (UK) Ltd. v. Mort Docks and Engineering Co. Ltd. (The Wagon Mound No. 1) (1961). Both are shipping cases:

The *Polemis* stevedores were unloading the ship *Polemis* when they carelessly dropped a plank of wood into the fume-laden hold. The plank struck a spark which in turn caused the fumes to explode and destroy the ship. The court held that the explosion was the *direct result* of the falling plank, and as the stevedores' carelessness had led *directly* to the destruction of the ship, they were held liable for the damage so caused. The dropping of the plank was bound to cause damage; the extent of the damage was not relevant to the question of liability.

The Wagon Mound. The ship was bunkering fuel oil in Sydney harbour, and spilled oil into the water. This spillage drifted across the harbour to docks which carried out marine repairs. The fuel oil was absorbed by floating cotton waste, and this in turn was ignited by sparks falling from welding operations. As a result, the docks were destroyed. The court held that because the destruction of the docks was *not a foreseeable* risk, there would be no liability for their destruction. Damages would only be awarded for the extent that could be expected for the fouling of the slipways.

Since the decision of the *Wagon Mound* case, the foreseeability test has been the one followed. It is unlikely that the directness test would ever be applied, but it remains a House of Lords decision and until it is removed from the English Law by statute, it must not be ignored.

A pilot who is negligent or careless will only be held liable for the resulting damage that he could reasonably foresee. Regrettably for the pilot, carelessness in an aircraft can foreseeably cause a great deal of damage if his error is likely to cause a crash. He can foresee injury or death to his passengers or fellow crew-members. He can foresee serious damage or destruction of the aircraft. If the aircraft is flying over a populated area, he can foresee damage and injury to the property and lives of those beneath him. If he crashed into the open sea, it may not be foreseeable that a ship was positioned at the point of impact, but it would be different if he were to crash into a harbour. The same would be true of a crash over deserted country. We can foresee the dwellers of populated areas, but not the unfortunate explorer or camel-mounted trekker of the desert who may be involved. Between the desert, and the open sea, and the cities, the particular facts will dictate the position, and a flight over Europe will carry more responsibility in this regard than an Atlantic crossing. Deviation from track is more hazardous over Europe than would be the case over the Indian Ocean; as witness the

tragedy over France in March 1973 involving two Spanish aircraft.*
The air collision over the Grand Canyon† in June 1956 in clear
weather, however, was less foreseeable. The airways were not so congested 20 years ago and the reliance on visual separation more widespread than at present.

The foreseeability test is applied according to the standards of the
defendant's knowledge at the time of the occurrence. In *Roe v. Minister
of Health* (1954), disinfectant, in which ampoules of anaesthetic were
stored, had seeped into the ampoules through visible cracks. The possibility that this might occur was not generally known at the time of the
incident which occurred in 1947. The plaintiff, who received a spinal
injection of the anaesthetic, became paralysed. The hospital authorities
were held not liable because the risk to the plaintiff was not reasonably
foreseeable at that date. 'We must not look at the 1947 accident with
1954 spectacles', said the Judge, Lord Denning. To take an example
from aviation history, consider the take-off of an Elizabethan aircraft,
G-ALZU, which crashed at Munich in 1958 (Chapter 1, pp. 49–53).
It is now known that the crash was due to the depth of slush that lay
on the runway, but at the time of that take-off, the effects of slush were
not generally known.‡ Unfortunately, for other reasons which are
discussed elsewhere in this book, the conduct of the enquiry and the
validity of its findings are now thought to be unsatisfactory. The pilot
did not receive justice, but the incident still serves as an excellent
illustration of using today's spectacles to look at yesterday's accident.
The pilot's good name was restored by a subsequent enquiry§ but too
late to save his professional career.

Another example is the crash of a BAC 1-11 during stall trials over
the Salisbury plains. The effect of deep stalls with high tailplane aircraft
was not known at the time. Fortunately the aircraft crashed on deserted
ground and no person other than the crew members was killed or
injured, but if anyone had been so unfortunate, it is unlikely that it
could have been shown on his behalf that the crash was foreseeable.
Today, however, the effects of slush and deep stalls are known, and
the pilot who repeats erstwhile blameless manoeuvres could well find

* Collision between Coronado and DC9 aircraft. See Chapter 1, pp. 57–9.
† Collision between DC7 and Constellation aircraft at 21 000 ft.
‡ A most unfortunate example: both Captain Price and Captain Bressey (p. 51) use this phrase
in all sincerity. See Chapter 3, pp. 104–5.
§ CAP 318; see Chapter 1, pp. 52–3.
‖ CAP 219, *Report on the accident to BAC One-Eleven G-ASHQ at Cratt Hill, near Chicklade, Wiltshire,
on 22nd October 1963.*

himself judged negligent. With the introduction of supersonic aircraft and galloping technology, this particular doctrine will be even more important.

Ingredient (3): proof of carelessness – breach of the duty of care. It is here that the definition of negligence given by Baron Alderson is most useful. The standard required is that of the reasonable man – in this context, the standard of the reasonable pilot. The 'reasonable man' has been judicially defined in a number of leading cases, all of which are worthy of note:

> The standard of foresight of the reasonable man eliminates the personal equation and is independent of the idiosyncrasies of the particular person whose conduct is in question. Some persons are by nature unduly timorous and imagine every path beset with lions. Others, of more robust temperament, fail to foresee or nonchalantly disregard even the most obvious dangers. The reasonable man is presumed to be free both from over-apprehension and from over-confidence.
> (per Lord Macmillan in *Glasgow Corporation v. Muir* (1943))

And as extracted from a number of cases:

> 'A reasonable man does not mean a paragon of circumspection.'
> 'He is cool and collected and remembers to take precautions for his own safety even in an emergency.'
> '...so on the one hand, an error of judgement may not amount to negligence. On the other hand the fact that "it might happen to anyone" is not necessarily a defence.'
> 'The test seems to be: what would be foreseen by a reasonable observer of the class whose conduct is in question?'
> 'The reasonable man is someone with a regard to caution such as a man of ordinary prudence would observe.'

From these definitions it is seen that a pilot must be judged for his reasonableness through the eyes of reasonable pilots. This point has been further examined by the courts in the case of *Bolam v. Friern Hospital Management Committee* (1957). Justice McNair said:

> Where you get a situation which involves the use of some special skill or competence, then the test as to whether there has been negligence or not is not the test of the man on the top of the Clapham Omnibus* because he

---

* A British colloquialism (particularly in Law), for the 'reasonable man'. Winfield and Jolowicz (*Tort*, 9th edn., Sweet and Maxwell, London, 1971, p. 26) add '...an American writer defines (him) as: "the man who takes the magazines at home, and in the evening pushes the lawn mower in his shirt sleeves"...'

has not got this special skill. A man need not possess the highest of expert skill; it is well established law that it is sufficient if he exercises the ordinary skill of an ordinary competent man exercising that particular art.

So the pilot who flies and conducts the operation of his aircraft with the skill to be expected from competent fellow pilots should not be found negligent if he makes an error of judgement. It may well be that some of the well-known aces in aviation could have recovered the situation, or would not have made the error; but that will not avail the plaintiff if he is unable to show that no reasonable pilot would have made the error of judgement in question. This application of the standard of skill demanded of the pilot was illustrated in an American case:

> *C. A. Colo 1950:* In the absence of Statute providing otherwise, pilot is not required to exercise extreme care and caution in operation of airplane, but only ordinary care in circumstances. 'Ordinary care' of airplane pilot is doing or failure to do that which an experienced pilot, having due regard for safety of himself and others, would do or fail to do under same or similar circumstances.
> (*Long v. Clinton Aviation Co.*)

Flying is steeped in technology and science. To a greater extent it is totally bound up in the nebulous art of airmanship, which is itself closely akin to seamanship. Pilots, like sailors, probably react differently to the same kind of critical situation and it is not for one to condemn the methods of the other. Both may probably be right. Events which subsequently occur may well lead to an emergency for one of the pilots in question but not for the other (see Fig. 11, p. 118). There are no grounds for condemning the pilot who finds himself in the emergency situation based only on the simple grounds that if he had followed the other's technique, no emergency would have arisen. A great deal more is required and this is one reason why there are few cases on this subject. There are other reasons which will be dealt with later.

Clerk and Lindsell on Torts, 13th edition,* puts the position in this way:

> The standard of the reasonable man varies with the circumstances. There is the balance of degree of likelihood that harm will occur on the one hand, and the cost and practicability of measures needed to avoid it, the gravity of the consequences, the end to be achieved by the conduct in question,

* Clerk, G. F. and Lindsell, W. H. B. *The Law of Torts*, 13th edn., Sweet and Maxwell, London, 1969.

including its importance and the social utility, and the exigencies of an emergency, dilemma or sport on the other.

An example can be found in the situation in which an aircraft approaches its destination airport in bad weather. The aircraft has a full load of passengers and adequate reserves of fuel. There *is* an alternative airport with satisfactory weather conditions, but the destination airport is the home base of the crew. They are familiar with that airport and have confidence in the landing aids to be used.

The commander and his crew members each, therefore, weigh up in their minds the problems posed by the position outlined above. These doubtless include:

(a) The likelihood of injury: there are times when flying an aircraft can be very taxing on the pilot's skill – but are there grounds for only flying when the conditions are easy? Professional pilots are expected to cope with difficult situations – a fact reflected in their training and in the frequency of their competency checks.

This pilot is very much aware that he has a difficult landing to make; but on the basis of his years of experience, and confidence in his own ability, he knows that with precision flying he can come down to his legal limits without danger. He feels alert and able to cope with the standards required of him.

His crew members also make their own assessment of the risk of injury, and, whilst some of them might not have attempted an approach had they been in command, they are confident of their captain's ability, and are prepared to assist him.

(b) Clearly, the way for the pilot to avoid the situation is for him to divert to his alternative airport where the problems do not exist. But the cost of aviation is phenomenal; a diversion to his alternative airport would involve an hour's flying, additional landing fees, and handling charges for an agent to look after his passengers and feed them while waiting for the weather to improve at his home base. In addition, the aircraft he is flying is probably wanted for the operation of another service out, and if the next service is cancelled there will be considerable loss of revenue. All this could well amount to many thousands of pounds.

There are other considerations. With the weather at the main airport down on limits, the alternatives are likely to become congested by other operators. The pilot may well have been on duty for some considerable time and if he does not get back to his base soon, will have to go off duty to avoid fatigue. This will add hours of delay. It is true that none of these considerations warrant risk to life and limb, but he has already assessed the risk and does not fear an accident.

(c) The gravity of the consequences should the risk materialise is all too clear. The crew members are well aware that lack of precision flying and skill can lead to an accident and the past has taught them that flying accidents have the most serious results. Aircraft of the present day are unable to fly so slowly that an accident results merely in a dented wing-tip; and for the aircraft in question to remain controllable, it must fly at a speed of not less than 130 m.p.h. The possible consequences of impact with the ground at that speed are not lost on the crew members.

And finally,

(d) The primary object of the flight was to get the passengers where they wanted to go, at the time they expected to be there. An airline with a bad reputation for keeping schedules tends to lose its passengers. At this stage there is no emergency or dilemma situation; there are reserves of fuel and the alternative airport's weather remains satisfactory.

The die is cast. The pilot elects to make an initial approach to assess the situation further. The combined efforts of his crew work well and the pilot is satisfied with the precision with which they work. The aircraft is stable and follows the landing aids accurately. Unfortunately, the fog is patchy and on the first approach when reaching critical height, the pilot is unable to make visual contact with the runway and cannot land. Automatically the crew go into the overshoot procedure and climb away to reconsider the situation. Adequate fuel remains; the first approach has been so satisfactory that the crew members have increased their confidence. The chance is good that on a second approach, the fog bank might have shifted slightly and visual contact will be made with the runway, so permitting them to land. However, the second attempt to land fails and the aircraft returns to a holding stack to await a change in the weather.

Following a reported improvement in the visuality, a *third* approach is made. This time, snatches of the approach lights are seen through the fog and with this promise of visual contact the aircraft continues down to a lower height than before. But the fog is deceptive, and visual contact is lost; an emergency situation has now arisen and great speed of reaction is required. Power is applied and the nose of the aircraft raised sharply; but because of the very low level of the aircraft and the speed required to execute the manœuvre, the rapid rotation of the nose subjects certain vital instruments to pressure fluctuations. A high rate of climb is indicated, but no acceleration in forward speed. The nose is eased down to try to increase the speed. Still the speed does not increase but the rate of climb indicator shows an adequate climb. To build up the inadequately low forward speed the nose is pushed further down. Although the aircraft is now in fact losing height, the pressure instruments show a rate of climb. The gyro-operated artificial horizon shows a marked nose down attitude, but the pilot is over-dependent on his inaccurate pressure instruments and, still chasing the airspeed, pushes the nose down even further. In the final moments, the error is recognised and an attempt is made to recover; but it is too late. The aircraft sinks to the ground and breaks up. Disaster has struck and amongst the burning wreckage, the passengers and crew are dead and dying.

Appraisal and debate. It is all too easy to say that the pilot was in error; that he should never have made the approach in the first place. To ensure the safety of aviation, no aircraft should fly. It is quite clear that the pilot must have made some error otherwise the crash would not have occurred. Perhaps the pilot should have taken more time to assess the situation? On closer analysis he might have realised that his vertical speed indicator was giving false information, although (for the purposes of this example) it should be stated that at the time of the crash, the effect of the pressure fluctuations described was not generally known to civil pilots. The evidence of the 'black box' flight recorder read-out and a careful enquiry will show the error that was made (see Fig. 12, p. 128); but can anyone say that the pilot of the unfortunate aircraft was negligent? He had but fractions of a second to assess a complicated situation and take remedial action. He knew of the risk involved but on the balance of considerations, he elected to make the fatal approach. If it can be said that no reasonable pilot would have done what this pilot did, then in law he is guilty of negligence. What concerns the reader is whether the pilot was acting in accordance with the standard of the reasonable pilot. If he is found unreasonable, then he is liable for the devastation that resulted. If he was not unreasonable, then society can only retire to lick its wounds and no liability is to be attached to the pilot. Compare the facts of this hypothetical case with the accident to Vanguard G-APEE at London Airport in 1965.*

A person must not disregard even remote risks. If he is aware that a risk is attached, then he must give it his consideration. The grounds at a cricket pitch were the subject of consideration in a leading British case. They were so arranged that a nearby public way was unlikely to be endangered by a well-hit 'sixer'. A particularly powerful batsman hit a ball on to the public way and caused injury. 'It does not follow that no matter what the circumstances may be, it is justifiable to neglect such a risk if he had some valid reason for doing so ... He would weigh the risk against the difficulty of eliminating it.' (per Lord Reid in *Bolton v. Stone* (1967)).

The question of whether the pilot is experienced or inexperienced is not necessarily a criterion in deciding whether he is negligent. If the pilot in question is a professional pilot, then the standard he is expected to show is that of the reasonable professional pilot. 'The principle is that a person who undertakes to do work which requires a special skill, holds himself out as having that skill; the lack of it then becomes blameworthy.' (*Philip v. Whiteley* (1938)). 'If you profess skill, then you must show the standard of those who normally do that kind of work'; 'It is not sufficient to be doing one's incompetent best.' These maxims can be illustrated by various examples. The pilot untrained in instrument flying who attempts to fly blind; the pilot whose only experience in navigation lies in following roads and railways, but who nevertheless attempts an ocean crossing; or a pilot trained only on single-engine aircraft who attempts to fly a complicated twin-engined aircraft. It may be seen from the following case that the issue may be more complex.

To use colloquial terms, most pilots occasionally 'thump' an aircraft into the ground instead of 'greasing it on' – that is to say, make a landing with less finesse than is good for the pilot's personal ego. Occasionally, however, for a variety of reasons, the landing is so 'firm' that

---

* CAP 270, *Report of the Public Enquiry into the causes and circumstances of the accident to Vanguard G-APEE which occurred at London (Heathrow) Airport on 27th October 1965.*

This aircraft also made three attempts to land in poor visibility, and crashed on the final attempt. Among the causes listed in the observations and recommendations of the Report are: 'pilot error due to low visibility; tiredness; anxiety; lack of experience of overshooting in fog; disorientation; over-reliance on pressure instruments; position error in pressure instruments', etc.

The Report also refers to the successful landing of another Vanguard (Echo Delta) after G-APEE's second attempt: '... At 0111, although there had been no improvement in weather conditions, Captain Shackell (pilot of G-APEE), probably stimulated by Echo Delta's success, asked permission to make another attempt to land ...'

damage or excessive stress to the airframe could result. Such damage may not manifest itself and could be difficult to find. In these incidents, a heavy-landing check is called for. This is not only embarrassing to the pilot concerned, but results in an expensive delay even if no damage is actually sustained. Questions are asked, and possibly the airline's training section will take a long, hard look at the pilot's record.

What differentiates a 'thumper' from a heavy landing is ill defined and really a matter of judgement for the technical crew of the aircraft concerned. But a responsible commander swallows his pride when in doubt and calls for a heavy-landing check. A less responsible man may convince himself and his crew that it was just 'rather firm' and ignore the possibility of hidden damage.

In June 1974 a Boeing 707 made such a heavy landing in Crete. An Accident Investigation Branch enquiry has published its report on the incident after investigating the facts which finally came to light. It seems that the heavy landing had been logged as 'very firm' but no steps were taken to carry out a heavy-landing check. 156 passengers were boarded and the aircraft flown back to Gatwick in the UK.

In fact, serious damage had occurred in the heavy landing which weakened the airframe; but fortunately the structure did not give way on the return flight. No one was injured; but if injury had followed, it would have given the aircraft commander some difficulty in defending an action in negligence.

The above-mentioned maxims do not apply to the situation when the person in question is compelled to act in an emergency or a dilemma. The rule in the case of *The Bywell Castle* (1879) is thus: 'The degree of care varies with the circumstances. A person who takes a reasonable decision as to a course of action in an emergency or dilemma, will not be treated as having acted negligently if the course of action decided upon turns out to have been the wrong one.' This rule extends beyond the fear of personal injury, and emergency action in the protection of property is also protected.

The *Bywell Castle* case concerned a ship in distress. The court held that ordinary skill and care are all that are expected of persons in charge of vessels; and not the ability to see at once the best possible course to pursue under the pressure of extreme peril brought about by the wrongful act of another.

It may be that the pilot of an aircraft is given wrongful or impossible instructions by the air traffic controllers with whom he is in contact.

The pilot may not have all the facts available to him to help him decide how best to deal with the situation; but if it subsequently becomes evident that he made the wrong decision under the pressure of his dilemma, then it does not necessarily follow that he was negligent.*

Ingredient (4): remoteness of damage. This is so bound up with foreseeability, that very little extra need be said about it. The principle from the *Wagon Mound* case applies. If the damage could be foreseen as a result of the accident, then liability will attach. There are, however, reasonable limitations to this rule. Aircraft carry senior executives travelling on company business. The loss of these men could well result in lost contracts and financial difficulties for the business concerns, but this loss is too remote and no liability will be awarded. Again, the aircraft may be carrying vital spares for some stranded ship. The vital spares may perish, and the owners of the ship be faced with considerable loss; but the damages are only likely to extend to the ship's owners. Each case will depend on its own merits.

   If it is established that the first four ingredients have been met, namely, (1) the pilot owes the plaintiff a duty of care; (2) the damage suffered was the foreseeable result of the wrongful action of the pilot; (3) the pilot was in breach of his duty to the plaintiff; and (4) the injury complained of was not too remote, then the court will find the pilot liable for his negligence. The next step is (5) to assess the extent of the damage attributed to the defendant, and finally, the monetary estimate of that extent of the damage.

Ingredient (5): the damage attributed. So far it has been assumed that the wrongdoings under discussion were the result of one pilot's error. It hardly needs saying that this is not always the case. Accidents occur through the combination of many factors, and sometimes, the actions of many persons. This topic will be confined to the question of contributory negligence.

   The law had an uncertain start, but briefly three situations can be examined: (a) where the plaintiff was partly to blame for the injuries caused; (b) where two persons are sued together by the injured party; and (c) where only one of the wrongdoers is sued.

---

* See text and report on DC9/Coronado aircraft collision, Chapter 1, pp. 59–60 in which one pilot found himself in this situation.

(a) When the plaintiff is partly to blame, there are statutory enactments that regulate the position. The Law Reform (Contributory Negligence) Act, 1945, s.1 states:

> Where any person suffers damage as the result partly of his own fault and partly of the fault of any other person, a claim in respect of that damage shall not be defeated by reason of the fault of the person suffering the damage; but the damages recoverable in respect thereof shall be reduced to such an extent as the court thinks just and equitable, having regard to the claimant's share in the responsibility for the damage.

The vast majority of persons injured by a pilot's error cannot be held to have contributed to their injury. There are, however, a number of situations where it could arise. For example, a fellow crew member could contribute to the accident or incident that brought about injury. The Commander, or Captain of the aircraft is generally regarded as being responsible for the safety of the flight. It is all too easy to see how an incompetent or negligent crew member can make the operation of the aircraft so much more difficult that situations are created which contribute to the pilot's error; take the case of a command to raise the flaps, given at the wrong speed, for example. The danger of doing this in modern high performance jets cannot be over-emphasised. The aircraft will stall and it may not be possible to recover from the stall before striking the ground (Chapter 1, p. 61). The primary duty of a co-pilot is to monitor the flight, assess any wrongful operation and if necessary query the command. If, when told to raise the flaps at the wrong speed, he unquestioningly does so, he must share the blame for any resulting accident.

Another example could be damage to aircraft involved in a near miss or collision when the navigation of both aircraft was in error.

(b) Where both wrongdoers are sued, then the court assesses the degree of responsibility that must be attached to each of the wrongdoers and the damages are awarded accordingly.

(c) When only one of the wrongdoers has been sued, then the defendant can claim contribution from his fellow wrongdoers. There is no bar to an action against only one of the persons who caused the injury.

It has been said that 'the plaintiff can choose his victim'. If the victim so chosen wishes, he can sue the other wrongdoer for a proportion of the damages awarded against him. Suppose the cause of an accident was a combination of wrongful air traffic control instruction, combined with careless navigation. The injured plaintiff can choose to sue only the pilot. The pilot, if damages are awarded against him, can claim contribution from the negligent air traffic controller.

Ingredient (6): the assessment of the damages attributed in monetary terms. This is a vast and complicated field of the law. In terms of quantum, it is not unusual for persons injured in motor accidents to have damages awarded in excess of £10 000. The value of an aircraft used in civil transport may cost over £2 m. If the aircraft contains over a hundred passengers, simple mathematics will show that the damages involved amount to an astronomical sum of money.*

Res ipsa loquitur

Where the laws of a country have the general rule that default is a *prerequisite* for liability, the plaintiff may be in some difficulty. The onus is upon him to prove the default of the defendant. He may have no difficulty in proving that an accident in fact took place, but he cannot prove how the accident happened, because the true cause of the accident lies solely within the knowledge of the defendant. The principle of *res ipsa loquitur* can avoid this hardship. The translation is: 'The thing speaks for itself', and Salmond, on Torts,† puts the matter thus: 'The maxim ... applies whenever it is so improbable that such an accident would have happened without the negligence of the defendant that a reasonable jury could find without further evidence that it was so caused.'

The leading case which illustrates the maxim is that of *Scott v. London and St Catherine Docks Co.* (1865). The plaintiff was an unfortunate passer-by who was struck by six bags of sugar falling from a warehouse. The Chief Justice, Sir William Erle said: 'There must be reasonable evidence of negligence; but where the thing is shown to be under the management of the defendant or his servants, and the accident is such as in the ordinary course of things does not happen if those who have the management use the proper care, it affords reasonable evidence, in the absence of explanation by the defendant, that the accident arose from want of care.' Firstly, therefore, it must be shown that the thing is under the management of the defendant or his servant; and secondly that the accident is such that in the ordinary course of things

* See pp. 246–7 for order of compensation envisaged in case of Turkish Airlines DC 10 crash, and for estimated liability of parties.
† Salmond, J. W. *The Law of Torts*, 15th edn., Sweet and Maxwell, London, 1969.

does not happen if those who have the management use proper care. It must also be observed that *when all the facts are known*, there is no room for the maxim, because the plaintiff has the evidence to prove negligence had there in fact been any. When the maxim does apply through *lack of knowledge*, then it affords reasonable evidence to the jury in the absence of explanation by the defendant.

The question that concerns the reader is whether the maxim can apply to aviation. Shawcross and Beaumont on Air Law* claim that the maxim does apply, and quote as their authority *Fosbroke Hobbs v. Airwork Ltd. and British American Air Services* (1937). In this case a plane crashed within 1000 yards of take-off. The court held that such a crash was not an inevitable peril of flying, and that the maxim should apply. It has applied on other occasions: in the case of a collision between a car and an aircraft; an aerial collision; and in respect of an aircraft which crashed while attempting a loop at between 200 and 400 ft altitude.

The maxim has been rejected by the courts, however, when a crash occurred following an engine failure on take-off; and in the case of the emergency landing and destruction of an aircraft when it was blown up by a bomb.

It is suggested here that only in the most obvious cases should the maxim of res ipsa loquitur be applied. Aviation is becoming so complex and technical, and acquiring so much new knowledge daily, that without this careful application great hardship and injustice may be inflicted on pilots and operators of aircraft. In their efforts to avoid liability, in fact, they will be under pressure to prove facts that are sometimes beyond the powers of government accident investigation agencies; for without such proof they remain vulnerable to Salmond's '... it is so improbable that such an accident would have happened without the negligence of the defendant ...'.

The law relating to res ipsa loquitur remains complex and uncertain. It will take the development of the doctrine by the high courts to clarify the uncertainty. This is still a grey area, and its application to aviation is one aspect of the law that will be worth following as and when cases occur.

Vicarious liability

The technical term of the relationship of the pilot and his employer

* Shawcross, C. N. and Beaumont, K. M. *Air Law*, 3rd edn., Butterworth, London, 1966. See also comment on 'air law', pp. 200–1.

is that of master and servant. The undertones of the pre-emancipated domestic servant of Victorian times illustrate the legal status, although the industrial environment of today has dispensed with the image of the humble worker. The pilot is a highly competent and responsible man; but his professional work is a service for the operator. The effect of this relationship means that as the service is for the benefit of the operator, the operator is responsible in law for the wrongdoings of the pilot.

A limitation on this broad statement is that the servant must not be on a frolic of his own. No one, of course, is employed under terms which permit him to carry out civil wrongs, and indeed, a contract in such terms would be void and of no effect. On the other hand, to say that because the employee committed a civil wrong he was *not* acting in the service of the employer is an evasion of the doctrine of vicarious liability, and will not be accepted by the courts. This applies even when the wrongful method has been expressly forbidden. The point was argued in *Century Insurance Co. Ltd. v. Northern Ireland Road Transport Board* (1942) in which case a fuel tanker driver set fire to a petrol station in the course of smoking a cigarette while dispensing petrol.

The reader may like to apply this test in connection with the breakup of a Boeing 707 in the region of Mount Fuji, Japan, in 1966. The aircraft was flying from Tokyo to Hong Kong. The normal route taken on these flights passes many miles to the east of Mount Fuji. The weather was clear and the captain elected to deviate from the normal route – probably so that his passengers could get a good view of the spectacle. The deviation from track was approved by the air traffic controllers. What was not known to the pilot, and had not been forecast for the flight, was that the weather conditions were ripe for particularly strong mountain waves around Fuji. When these waves exist, severe turbulence can be experienced. The unfortunate aircraft encountered abnormally severe turbulence and suffered structural failure, causing it to break up. All aboard perished.

It is submitted that the purpose of the deviation was to give the benefit of the spectacle to the passengers, thus fostering goodwill and encouraging them to fly with the airline again. The pilot was not on a frolic of his own.

While the particulars of this case are before the reader, he may like to consider whether the pilot was in any breach of duty to his passengers. It is submitted that there was no breach of duty. There was no lack

of airmanship. The only risk of flying outside controlled airspace is
that of encountering conflicting traffic, without the warning that the
air traffic control can give. But many flights take place outside controlled airspace, on regular routes. Even if the deviation had not been
approved by the operators, this would only involve the pilot in breach
of his contract. Had it been that the employers had stated that there
should be no deviation from track because of the *known* risk of severe
turbulence, then this might be another matter. These unusual conditions were unknown to the operator.

The master and servant relationship can exist in the world of the
private pilot, since this relationship does not necessarily involve formal
contracts, or the payment of money. Two examples can be
considered:

The owner of an aircraft travels, by other means, to the south of
France. He makes arrangements for a friend to fly his aircraft out to
him so that he can enjoy both the company of his friend and the use
of the aircraft. It is their intention to fly home together. On the out-
ward flight, the aircraft crashes due to the negligence of the friend.

The second example is the loan of a private aircraft to the employee
of the owner. The employee uses the aircraft to fly out to Spain on his
holiday. Due to his negligence, the aircraft crashes. In both cases, other
persons are hurt.

To assess the position, two motoring cases are considered:

Britt v. Galmoye (1928): A master lent his car to his servant to enable
him to take his friends to the theatre after the day's work. Due to the
servant's negligence, a third person was injured and sued the master
for the wrong of the servant. The master was held not to be vicariously liable, for in the context of the case, the servant was a mere
bailee of the vehicle using it for his own purpose, and not in the course
of his employment.

Ormrod v. Crosville Motor Services Ltd (1953): The owner of a Healey
car was held liable for the negligence of a friend who was driving it
from Birkenhead to Monte Carlo, for it was proposed that on arrival
they should use it for a joint holiday. Lord Denning said: 'It has often
been supposed that the owner of a vehicle is only liable for the negligence of the driver if that driver is his servant acting in the course of
his employment. That is not correct. The owner is also liable if the
driver is his agent, that is to say, if the driver is, with the owner's
consent, driving the car on the owner's business or for the owner's
purpose ... The owner only escapes liability when he lends it or hires

it to a third person to be used for a purpose in which the owner has no interest.'

Suing the pilot. The importance of the doctrine of vicarious liability to both the pilot and the injured person is that as the operators or, to be technical, the masters, usually have the funds behind them, they are better able to meet the damages that may be awarded. If an action were to be brought against the pilot alone it is possible that his assets would not be sufficient to compensate the plaintiffs. There might, however, be a situation whereby the owners of the aircraft were on the point of bankruptcy. A multiple award for heavy damages could prove to be too much for the company, and send it into liquidation – and thus, the successful claimant could find himself in line with other creditors struggling for a proportion of what he was owed.

However, the pilot whose negligence caused the accident may have sufficient funds to meet a particular plaintiff's claim. In this situation, the injured person might well be advised to sue the pilot and not the company. This would be a matter for careful consideration, and possibly, careful investigation.

An air crash occurred on 6 May 1962. A DC 3, G-AGZB, was operating a scheduled service from Jersey to Portsmouth. Low cloud and drizzle prevailed in the Portsmouth–Isle of Wight area, and when at approximately mid-channel, the aircraft notified the London Flight Information Region (FIR) that it was descending to 1000 ft. Shortly before the accident it was seen flying low over Ventnor towards the cloud-enshrouded St Boniface Down on which it crashed and burst into flames. Both pilots and eight passengers were killed instantly. The stewardess and another passenger subsequently died of their injuries.

The investigating officers' opinion was: 'As a result of an error of airmanship, the aircraft was flown below a safe altitude in bad weather conditions and struck cloud-covered high ground.' Regrettably, the case does not appear to have been reported in the usual law reports; but in this instance, *the pilot's estate was sued directly*. Thus the vicarious liability of the master does not absolve the servant from his own wrongs.

Fatigue and pilot error*

The law and its relationship to fatigue and pilot error is ill-defined. It cannot truly be categorised into either civil or criminal law although

* Fatigue is also discussed in Chapter 1, pp. 27–33, Chapter 2, pp. 86–7 and Chapter 5, pp. 187–91.

elements of both may be present. For convenience and clarity, it is appropriate that this subject should be attached to the question of negligence.

It will be useful, too, to restate the BALPA definition of the condition: 'That degree of tiredness which leads to an impaired ability to fly accurately and to make correct decisions'. (See Chapter 1, p. 29.)

The pilot who allows himself to become fatigued as defined above must be aware that he is increasing his chances of making an error. The consequences of making an error have already been discussed; but the law must examine whether the error was a simple misjudgement on the part of the pilot, or whether his action was unreasonable. A misjudgement does not normally involve liability. If the other necessary ingredients are present, an unreasonable action may well result in liability. When the question of possible pilot error arises during an investigation, the pilot may attempt to defend himself by claiming that the long duty day, or other circumstances which gave rise to the fatigue, made his error a reasonable one. Should the enquiry be conducted by the employers for the purpose of disciplinary action, and the cause of fatigue be plainly due to an extended duty period for the company's commercial interest, then the employers would be embarrassed not to take the fatigue into consideration. If the enquiry was the subject of litigation, however, then the lawyer's response could be to question whether or not the pilot was reasonable in allowing himself to become so fatigued in the first place – this in itself being negligent. That there do not appear to be any legal precedents in support of this argument is not surprising. The study of pilot fatigue is, after all, a recent science, notwithstanding that pilots have complained of its burdens for many years.

The judgement whether the pilot was reasonable in allowing himself to become fatigued is beset with difficulties. The pilot involved in such incidents would normally be the employee of a company. Those who fly for pleasure are unlikely to do so when struggling with the misery of fatigue; however, if these private pilots do fly for some compelling reason, then the same principles apply. For the professional pilot the dilemma lies between the demands of commercial considerations and the duty of getting his passengers to their destination with the minimum of delay, and the real difficulty of forecasting the extent of his fatigue at the end of a flight – that crucial period when it is imperative that he should not be so tired that the safety of the aircraft is endangered.

With hindsight or in isolation it is all too easy to say that safety

comes first. Safety does come first, but it is a very expensive com-
modity and there is always some degree of compromise.* Where that
compromise lies is a matter of judgement, and managements of the
airlines probably allow their judgement to sway closer to the thinking
of their accountants. The pilot is less influenced by the accountants;
but he carries the final responsibility for the commercial policies of the
managers and can run considerable personal risk in opposing their
policies. In some airlines, a Captain who decides to make an un-
scheduled night-stop because of pilot fatigue is courting difficulties with
his future career.

There are obvious limits to a claim that the dilemma of the pilots
justifies them in accepting long duty hours. Fear of victimisation does
not override the compelling duty of safety. If the duty is obviously
unacceptable and the pilot knows that he *will* be fatigued, then he
runs a serious risk in starting the flight. Yet the other horn of the
dilemma has been stated; that at the beginning of these long duty
periods it is very difficult for the pilot to see his operational ability
in the final stages. Pilots are not deemed to have the qualifications
of medical men, and consequently are not expected to exercise the
doctor's technical standard of judgement when assessing their degree
of fatigue. They do, however, have experience of their own personal
limitations.

The customs of a profession are very often used as the yard-stick
in assessing whether or not a professional man has shown the standards
of his colleagues. It is particularly difficult for one crew member to
refuse to fly when all his colleagues readily accept the duty. The fact
that a certain mode of action is an accepted standard, however, is not
a guarantee that it excludes negligence. 'Neglect of duty does not cease
by the repetition to be neglect of duty' (*Carpenter Co. v. British Mutual
Banking Co.* (1937)).

Some further guidance on this topic may be gleaned from the
criminal defence of necessity discussed later in this chapter. The solution to this problem lies in statutory controls with criminal sanctions
against the pilots and the operators. Most nations already have such
statutes, e.g. the British Air Navigation Order (see pp. 237–9). The
question as to whether the statutes are adequate or not is another matter
and the subject of fierce industrial dispute. This dispute is outside the
scope of this chapter, but very well illustrated in the BALPA committee report referred to above.

* See Designer's thought-processes in arriving at compromise. Chapter 3, pp. 96–7, 112–14.

Particular statutes

There are three statutes which are of vital interest to the question of pilot error and the law: (1) The Law Reform (Miscellaneous Provisions) Act 1934; (2) The Fatal Accidents Acts 1846 to 1959, and (3) The Carriage by Air Act 1961.

(1) *The Law Reform (Miscellaneous Provisions) Act 1934.* The Old Common Law Rule was 'Actio personalis moritur cum persona' – Personal actions die with the person. Thus, if a person was killed, his estate could not sue for his death on his behalf, and nor could it be sued. By virtue of this act, there is a survival of the rights of action, for Section 1 (1) provides: 'On the death of any person ... all causes of action subsisting against or vested in him shall survive against, or as the case may be, for the benefit of, his estate.'

The result of this is that the executors of persons killed can sue the pilot if he is alive, or the pilot's executors if he is dead. *In view of the large sums of money involved in aviation accidents, this is a very important provision, as the matter does not end for the pilot on his death. His dependants may well be left destitute if any person injured sues the pilot.* Article 27 of the Warsaw Convention also enables actions to be brought against the estate of persons liable for the damage caused. Where the Convention applies, it has the force of law by virtue of the Carriage by Air Act.

(2) *The Fatal Accidents Acts 1846 to 1959.* This act provided a right of action for classes of persons who are closely related to the dead victims. It does not put a price on the head of the unfortunate deceased as was done by the ancient 'wergild', nor does it take into calculation 'solatium' – the broken hearts of the relatives. The Act 'is an act for compensating the families of persons killed; not for solacing their wounded feelings.' (per Coleridge, J. in *Blake v. Midland Rail Co.* (1852).)

There are three requirements: (a) There must be a right of action such as would (if death had not ensued) have entitled the party injured to maintain an action and recover damages in respect thereof. If by some contractual provision in a ticket, the passenger had relinquished his rights of action, the relatives would also be deprived of their course of action. (b) The relatives must be of a limited class of dependants. This includes the wife, husband, children and grandchildren, also his or her brother, sister, uncle or aunt and their issue. It will be noted how this can affect the wide number of persons who can bring an action. (c) The basis of compensation is that of compensation for the financial loss of their breadwinner. In this context, it is most relevant to consider the degree of dependence on the dead victim.

(3) *The Carriage by Air Act 1961.* It is within this act that will be seen the probable reason why there is so little case law on this subject. The Act embodies the Warsaw Convention 1929 into the English law, and there are two vital provisions. The first is that where the Act applies, the operators of the aircraft are strictly liable for the death of a passenger on board an aircraft or in the process of embarkation or disembarkation thereof (Sch. 1 Art. 17). When compared with the enormous difficulty of proving negligence, as would be required if the pilot was to be sued, there is little wonder that the victims choose to sue the operator. The strictness of this provision can be illustrated by the incident when, on 6 October 1970, a Pan-American Jumbo jet was hijacked and diverted to Cairo. The airline and the crew were powerless to prevent the crime. Explosives were loaded on board by the hijackers and when the opportunity subsequently came to evacuate the aircraft, the emergency chutes were used. A number of passengers were injured and, although there was no question of negligence, they successfully claimed damages from the blameless airline.

The second provision is the limitation of liability that is imposed on the plaintiff. This is the equivalent of 250 000 francs* (Art. 20(1)) and is the total that can be claimed for the death of a passenger whether his executors sue the pilot, the airline, or both. The pilot may care to reflect that even with the protection of the Act, if by his negligence he kills 100 passengers, the damages that will be awarded against him for their death will not exceed 25 000 000 francs. That is little protection indeed for a pilot, but it may save an airline from bankruptcy. From the case of *Preston v. Hunting Air Transport Ltd* (1956) the damages under the Act appear not to be limited to financial loss. The Act does not apply to persons who are not passengers.

The law of contract

The law of contract deals with those persons who bind themselves by agreement. The laws of different countries differ enormously. One thing appears to be held in common, however, and that is that contracts are not necessarily formal documents which require ceremony,

* See p. 226 n.

signature or even writing. The ideal is for persons to make their agreements freely and include what terms are mutually acceptable. In aviation, this freedom is severely restricted by many statutes and international conventions, and where these apply, the normal law of the country prevails.

The law of contract as it applies to pilot error is restricted. The pertinent contracts are (1) Between the passenger, or shipper of cargo, and the operator. (2) Between the pilot and the operator. (3) Between the insurer and the insured of aviation risks and (4) between the lessor and lessee of aircraft.

(1) *Passengers, or shippers of cargo, and the operator*

In general the passenger can fly only by virtue of his ticket, and the shipper of cargo can send his goods only under an air bill. These are in fact written contracts, and include the terms of transportation which must be accepted by both parties. The standard International Air Transport Association ticket states:

> Condition 2: 'Carriage hereunder is subject to the rules and limitations relating to liability established by the Warsaw Convention unless such carriage is not international carriage as defined by that Convention.'

A convention is not a legal document unless ratified and embodied in a statute. For example, the English Carriage by Air Act 1961 gives the Warsaw Convention the force of law in England. The Convention applies to all international flights except those which transport between a place within the United States or Canada, and any place outside those countries. In these cases special tariffs in force in those countries apply. The Carriage by Air Act (Application of Provisions) Order 1967, SI 480, makes similar provisions for non-international flights within the UK.

The situation then is that the passenger and shipper of cargo is bound to accept the limitation of liability as laid down by the Convention. Any damages for death or personal injury and loss of or damage to baggage will be governed by the Convention.

Article 22 of the Convention limits the measure of damages for death or personal injury of each passenger to 250 000 francs. Passengers' hand baggage is limited to 5000 francs, and cargo is limited to 250 francs per kilogram, or its declared value.* The award of Court costs is addi-

---

* These figures are valid at the time of writing.

tional to the limited sums: for full details reference should be made to the Convention.

> Condition 3: To the extent not in conflict with the foregoing ... the services performed by each carrier are subject to ... carrier's conditions of carriage and related regulations which are made part hereof (and are available on application at the offices of the carrier) ...

The significance of this is that if the Convention or other regulating statutes do not apply, then the carrier's conditions of carriage govern the contract. What then if these conditions seriously limit or exclude liability for damages? Because the passenger's attention is drawn to the existence of these extra conditions and because they are available for inspection, they form part of the agreement even though the passenger concerned did not take the trouble to inspect them at the carrier's office. This point was clearly made in the case of *Thompson v. L.M.S. Railway Company* (1930). In this case an illiterate person was held bound by the terms of a ticket which was marked on the front, 'for conditions see back'. On the back was a notice that the ticket was issued subject to the conditions in the company's timetables and excursion bills. The court disregarded the fact that the person was illiterate on the grounds that 'illiteracy is not a privilege, but a misfortune'.

This principle also applies when one national buys a ticket written in a language he does not understand. No doubt all the big commercial operators of public transport have only the most reasonable terms that apply; but pilot error is not limited to big commercial operations. It could well be that a very small organisation has very restricted terms governing liability, or perhaps even excludes liability altogether. What then is the position of the injured passenger?

As explained in the section on tort the servant of the company is liable for his own wrongs. If the pilot error was actionable and caused injury, there is the question of whether the passenger could sue the pilot.

Condition 6 of the ticket states: 'Any exclusion or limitation of liability of carrier shall apply to, and be for the benefit of, agents, servants and representatives of carrier and any person whose aircraft is used by carrier for carriage and its agents, servants and representatives.'

Let it be initially stated that such a term is binding by virtue of Article 25A(1) of the Warsaw Convention and its ratifying statutes. Where the Convention applies, the term also applies, and the same limitation or liability protects the pilot. The problem posed is when the Convention does *not* apply and the carrier's own limiting terms make it desirable for the injured passenger or shipper of cargo to claim his compensation elsewhere. Article 25A(3) withdraws the protection if the damage was done with intent to cause harm, or recklessly with knowledge that damage will result. Can the pilot claim protection from Condition 6 of the passenger's ticket?

Within the English law we find the doctrine of 'privity of contract'. This means that the parties to a valid contract obtain and incur reciprocal rights and obligations. It is only the parties to a contract who can sue on it. Rights cannot be conferred on a stranger to a contract as a right to enforce the contract in person (*Dunlop v. Selfridge* (1915)). An exception to this rule is when a statute specifically excludes the doctrine, and the ratified Warsaw Convention in fact does this. Outside that, the doctrine of 'privity of contract' prevails, and it is therefore pertinent to examine it.

Three cases are worthy of note: *Adler v. Dickson* (1955), *Genya v. Matthews* (1965), and *Gore v. Van der Lann* (1967).

Adler's case. The plaintiff was a passenger in the Peninsular and Oriental Steam Navigation Company's vessel Himalaya, and was travelling on a first class ticket. This ticket was a lengthy printed document containing terms exempting the company from liability. A general clause that 'passengers are carried at passengers' entire risk' was reinforced by a particular clause that 'the company will not be responsible for any injury whatsoever to the person of any passenger arising from, or occasioned by, the negligence of the company's servants'.

While the plaintiff was mounting a gangway, it moved and fell, and she was thrown on to the wharf from a height of 16 ft, sustaining serious injuries as a result. She brought an action for negligence, not against the company, but against the master and boatswain of the ship.

The court held that while the clauses protected the company from liability, they could avail no one else. The master and the boatswain were not parties to the contract and could not claim its advantages. Lord Justice Jenkins pointed out that the contract purported to exclude only the liability of the company and not that of their servants; but, he continued, 'even if these provisions had contained words purporting to exclude the liability of the company's servants, non constat (it does not follow) that the company's servants could successfully rely on that exclusion . . . for the company's servants are not parties to the contract'.

Genya's case. An old age pensioner signed a written application form in order to obtain from Liverpool Corporation a free pass to travel on their buses. The form contained the following clause:

In consideration of my being granted a free pass for use on the buses of Liverpool Corporation I undertake and agree that the use of such pass by me shall be subject to the conditions overleaf, which have been read to or by me prior to signing.

One such condition stated:

The pass is issued and accepted on the understanding that it merely constitutes and grants a licence to the holder to travel on the Liverpool Corporation's buses subject to the condition that neither the Liverpool Corporation nor any of their servants or agents responsible for the driving, management or working of their bus system, are to be liable to the holder ... for loss of life, injury or delay or other loss or damage to property however caused.

Other conditions excluded the liability of the Corporation itself from a claim for damages.

The plaintiff was injured due to the negligence of the driver of a bus in which she was travelling, and she sued the driver. The judge at the hearing held that the driver was not a party to the agreement and could not claim its protection. The court felt that the agreement was only a licence.

Gore's case. This was not identical to Genya's, but the facts were similar. Again Gore was an old age pensioner who obtained a free pass on the Liverpool Corporation's buses and under the same conditions as in Genya's case. She was injured by the conductor's negligence. The Court of Appeal held that the agreement was in fact a contract and not a mere licence. Because of Section 151 of the Road Traffic Act, any clause for exclusion of liability for injuries in a public transport vehicle were void. The plaintiff was able to claim for her injuries.

There does not appear to be any reason why these principles should not apply to contracts (or licences) to fly in aircraft. An aircraft, like the Liverpool Corporation bus and the P & O ship, is only a form of transport. To alter the Common Law, there must be an Act of Parliament couched in clear terms. Reference has already been made to the two relevant British statutes. If they do not apply to a case under consideration, then the Common Law decides the issue.

(2) *Between the pilot and the operator*

The terms of employment of the pilot again govern the relationship of the two parties. The technical relationship is that of master and servant and, as explained in the section on tort, the servant is under a common law obligation to take care of his master's property, namely the aircraft, and to exercise a reasonable standard of skill in the execution of his duties.

If express contractual terms are absent, there is an implied condition that the servant should 'serve the master with good faith and fidelity and that he would use reasonable care and skill in the performance of his duties' (per the House of Lords in *Lister v. Romford Ice & Cold Storage Co.* (1957)). The facts of this case were thus: The appellant was employed by the respondents as a lorry driver. His father was his mate. While backing his lorry, Lister drove negligently and injured his father. The father sued the respondents, who were held vicariously liable for the son's negligence. The respondents now sued the son, inter alia, for breach of contract. The House of Lords gave judgement for the employers.

It can be seen how this decision applies in the case where one crew member sues the operator for the negligence of a pilot who caused his injuries. It is important to remember that the limitations to awards of damages restricting a passenger's claim do not apply to fellow crew members, as they do not fly by virtue of a ticket. It would be otherwise if the injured crew member was 'dead heading'* and travelling with a ticket, even though such tickets are free to the holder.

The terms of employment may expressly *exclude* this implied condition; this is the case with one American operator's contract of employment. If the pilot is sued by a third party for damage arising from the execution of his duties, his employers indemnify the pilot for the damages awarded. Whether such terms are desirable is a matter for speculation outside the scope of this chapter; but it must be noted that the legal study group of the International Federation of Air Line Pilots have advocated that all contracts of pilot employment should include such a clause. It is pertinent to recall that this book begins with the observation that 'the pilot is the first person to arrive at the

---

* Dead heading: 'The practice of transferring crews from place to place as passengers in surface or air transport either before or after a flying duty period.' (Definition from CAA publication.)

scene of an aircraft accident' and that this fact alone should be sufficient to ensure a high professional standard. The possibility of 'legal repercussions' is not normally uppermost in the pilot's mind when the question of 'pilot error' arises.

(3) *Between the insurer and the insured*

Reduced to its simple terms, a contract of insurance is an agreement between parties that, in return for a sum of money called the premium, the insurer indemnifies the insured against a particular loss, or against a particular happening. It is in essence a pure contract, and unless the legislature dictates otherwise, the basic law of contract prevails.

The actual insurance can take a number of forms. A person can insure himself against death or physical injury; he can insure his property against loss or damage, and he can insure himself against third party risks. This is much the same as motor insurance with which so many persons are familiar. The distinguishing feature of aviation insurance is the vast sums of money involved.* The aircraft are costly and the persons involved in major claims are frequently numerous.

Motor accidents occur all too often, and as the private motorist is usually unable to meet his compensatory liabilities, the legislature of many countries impose a compulsory third-party liability insurance as a condition of the use of a motor vehicle on the road. The question of compulsory insurance in aviation, however, has had something of a chequered career. The convention on 'Damage Caused by Foreign Aircraft to Third Parties on the Ground or Water', known as the Rome Convention, was first held in 1933; but it was not until 4 February 1958 that a reconvention obtained sufficient signatures to enable it to come into force. The convention put an onus on the operator of 'an aircraft registered in a contracting state, to insure to a prescribed limit against liability for death or injury to persons or damage to property on the ground or water caused by an aircraft in flight or by an object falling from such an aircraft'. Eighteen states including Australia and Canada ratified the convention. The UK signed the convention but did not ratify, and the USA did not sign.

The UK did not ratify because it considered the limitation of liability too low and would therefore cause injustice to the persons suffering loss. Section 40 of the Civil Aviation Act 1949 imposed liability on aircraft operators for damage commensurate with the Rome Convention, but not a strict liability: proof of negligence was required.

* See p. 246 n.

In preparation for a possible ratification of the Rome Convention, sections 42 to 48 of that Act made provisions whereby compulsory insurance could be enforced by an Order by the responsible Minister. Unfortunately the ratification never came about and sections 42 to 48 were eventually repealed by Britain's 'Companies Act' of 1967.

In the absence of a truly international agreement, compulsory insurance in aviation may apply only to a limited number of states. Where there is no insurance, it is possible that the operator or pilot must look to his own pocket to match his liabilities – not a happy state of affairs. What is probable, however, is that although a legislature might make insurance compulsory, the aviation authorities may refuse to grant an air operators' certificate unless adequate insurance is taken out. Cautious passengers might like to question the operators on this point before travelling. Persons on the ground are not so fortunate and are not able to question whether or not the operators of aircraft flying overhead have adequate insurance cover.

In cases where there *is* insurance, little difficulty arises. If the insured has taken out a policy on his own person and he is killed or injured, either he, or his personal representatives, claims against the insurer. The same applies if a person insures his freight or personal effects. If they are damaged or lost, he claims on the insurance company. If the insurance company refuses to pay up in breach of their agreement, then he can sue in contract.

Third party liability is not so straightforward. Firstly, the injured person will have to establish a claim against the wrongdoer. This will be in negligence, breach of statutory duty, or strict liability as already discussed under tort. It is only when the person sued is found liable that his insurance cover takes effect to indemnify him against the claim. If the person can escape liability for his actions, then no matter what insurance cover he holds, his insurer has nothing to indemnify.

This position was expressed in the English motoring case of *Bennett v. Tugwell* (1971). Tugwell was the driver of a car insured against risk to passengers. He affixed a notice to the dashboard stating 'Passengers travelling in this vehicle do so at their own risk', in the mistaken belief that in the event of an accident, the passenger could only sue the insurance company and not the car owner or driver. Bennett, knowing of the notice, travelled as a passenger in the car and suffered injury due to the admitted negligence of the driver. Justice Ackner pointed out that since a policy of insurance was a contract of indemnity, the insurers were under no liability to make any payment under that contract if their assured is himself under no liability. Bennett had accepted the risk and could not claim. However, the Judge went on to express stringent criticism of insurance companies who accepted premiums against liability, and then chose to avoid that liability by professing that the passenger had voluntarily accepted the risk.

In England, the Motor Vehicles (Passenger Insurance) Act 1971 has altered the position of the passenger in road vehicles, but the indemnity principle of insurance remains. If the wrongdoer is not liable, then his insurance company has nothing to pay. As stated earlier in this chapter, the majority of airline passenger tickets are bound by the Warsaw Convention which imposes a strict liability on the operators in relation to their passengers. If the operator is insured, the liability can be passed on to the insurance company. When the convention does *not* apply – for example when the injured person is not a ticket holder – then the principle of indemnity prevails.

Certain countries, notably England, have the doctrine of Privity of Contract as discussed earlier. The insurance contract is between the insured and the insurer. Unless statute provides otherwise, it is only the assured who can sue the insurer should he fail to pay up. It is difficult to foresee problems in this regard when the aircraft is being operated by a commercial organisation. There is little difficulty in tracing the operator and good business dealings ensure that the protection of the insurance cover is forthcoming. Even in the event of bankruptcy or other insolvency of the operator – possibly due to the crash – the injured party may still be able to claim from the insurance company by virtue of the Third Parties (Rights against Insurers) Act 1930. This Act gives only partial assistance because the claim is subject to any limitations contained in the policy; thus, if for some reason the assured was prevented from claiming, then the third party is equally restrained.

Should the owner or pilot of the aircraft be a private individual he could possibly make a clandestine departure beyond the reach of the courts. This could cause serious complications and the injured person may find that he has no remedy, even against the insurance company.

The Law of Aviation Insurance is a subject more complex than can be adequately expressed in a short section of this chapter; however, it is felt that sufficient indication has been given here of the maturity of this field.

(4) *Between the lessor and lessee of aircraft*

The fundamental point for consideration is whether the lessee of the aircraft is acting as bailee, or servant or agent of the lessor. This can only be ascertained from the terms of their contract:

(a) If the relationship is that of bailee, then the bailor will *not* be liable for the wrongdoings of the bailee or his servants. An American case illustrates the point:

> D.C.N.Y. 1954. Owner who rented to a suitably licensed and qualified pilot an airplane which, at the time of rental, was airworthy and properly maintained, inspected, tested and equipped in all respects, and who did not operate or control the craft at time of accident, was not liable for injuries sustained by third parties when the airplane crashed while under control of renting pilot. *D'Aquilla v. Pryor.*

(b) If the relationship is that of master and servant or agent, then liability is attached to the lessor; again an American case illustrates:

> C. A. Miss 1955. Owner who authorised pilot to use his plane becomes liable for negligence of pilot in operation of plane. Code Miss. 1942. 7533 to 7544–17. *Hays v. Morgan.*

The situation becomes more complicated in the case of a 'wet lease', this being a situation whereby a company leases its aircraft and an operating crew to another airline. This is frequently done by the big operators when supporting their subsidiary or satellite companies. If the wet-lease pilot makes an error and damages arise, who, in law, is the master of the wrongdoing servant? The question was examined in the case of *Mersey Docks & Harbour Board v. Coggins & Griffith (Liverpool) Ltd.* This is a House of Lords hearing and of the highest authority in England.

The Mersey Docks board owned a number of mobile cranes which were let out on hire to stevedores to enable them to load and unload ships. The terms of hire were that the hirer 'must take all risks in connection with the matter ... (and) the drivers provided with the cranes shall be the servants of the applicants'. The driver of the crane provided in this case was negligent and injured a third party. At the time of the accident, the stevedores could give operational directions to the crane driver, e.g. what pieces of cargo were to be lifted, but had no power to direct the driver in the manipulation of the controls of the crane. It was held that the owners of the crane were liable for the injuries to the third party. The question will often be asked as to *who* has the right to control the way in which the work is to be done, and the answer to this question may also answer the problem: but it is not the universal test of the master–servant relationship. The fact that the master does not have the technical knowledge to do the job or give the orders is irrelevant, and the peculiar skill of the pilot makes him no exception to this rule. Authority for this statement comes from the many cases in which hospital authorities have been held liable for the negligence of their doctors. The only way to establish who is the master of the servant is to take all the facts into account.

Pilot and passenger

Under the law of contract, there is a fifth form of relationship that should be examined in a negative sense, namely, the relationship between the pilot and the passenger. Aside from the very small organisations, the normal situation is that both the pilot and the passenger have contracts with the operator of the aircraft, but they have no contractual relationship between themselves. There is no privity of contract. Within the law of tort, the pilot owes his passengers a duty of care, but breach of this duty is not actionable at the suit of the passenger unless he has suffered damages.

In March 1973, the Air Traffic Control Network of France was disrupted by the industrial action of the controllers, and French Military air traffic control officers took over these duties (Chapter 1, pp. 57–8). Considerable disquiet was felt both by the operators and the pilots of aircraft flying services over France during this period. The French Government claimed that the military controllers were most competent and that there was no lowering of safety standards – a view which a number of airlines and pilot associations did not share. Their pilots either refused to fly over France until the normal control was resumed, or accepted that there may have been some change in standard, but that it did not warrant the commercial considerations of suspending services. This was entirely a matter of judgement; the pilot, by virtue of his natural characteristics, avoids dangerous situations, but he is also very alive to the commercial considerations of his operators, and accepts a calculated risk in pursuit of his duties.

The question arises: 'Does a passenger have a right of action against the pilot for flying in a hazardous situation, even though he has suffered no injury or damage?' The answer seems to be that as he has suffered no damage, he has no claim in tort. Can he then sue under the implied terms of his contract that he is entitled to the safe operation of the aircraft in which he is flying? Against the pilot, he has no contract, therefore it follows that no such term can apply. His contract is with the operator. There does not appear to be any judicial decision whether or not such an implied term exists, but, even assuming that one could be proved, the passenger would have considerable difficulty in pitting his judgement against that of an experienced airline. The criminal aspects of such a case are discussed later, but the committing of a crime, in itself, is no ground for a civil suit.

During this period of disrupted control over France, two Spanish-operated aircraft, a Coronado 880 and a DC9 elected to fly into the French air lanes. The operators and the pilots were aware of the situation. In their judgement the control facilities over France did not warrant cancellation or division of their services. It could be that the pilots were in the dilemma situation as described in the section on fatigue and that, while they felt disquiet, they did not refuse to operate. This is mere speculation and it is doubted whether the pilots' personal feelings or opinions will ever be established, but for the sake of illustrating the point, let it be assumed that the pilots were not content. The tragic fate of the Coronado 880 and the DC9 was that they collided with each other. This chapter has already dealt with the legal problems that follow an accident. It suffices to conclude that if they had *not* collided, it is very unlikely that the passengers could have succeeded in any claim against the pilots, simply because the pilots took a risk.

Criminal law

As the subject of this chapter is pilot error, there is no need to consider the legal complexities of criminal intent, or the question of malice. In contrast to the civil law, which is almost pure common law and applicable to the general public, including pilots, the criminal law as it relates to the subject of pilot error is applicable only to pilots, and the operators of the aircraft flown. A simple parallel is the embodiment of motoring law into the Road Traffic Act. It is unlikely that a pedestrian will have his shoes impounded for jay-walking. It is equally unlikely that anyone other than the pilot or operator will become involved as a defendant in a criminal prosecution for pilot error. As the Road Traffic Act applies to the motorist, the Air Navigation Order applies to the pilot and operator.

This is not to say that a pilot will not be subject to a criminal prosecution for an offence outside the specialised statutes. Article 43 of the British Air Navigation Order states: 'A person shall not wilfully or negligently act in a manner likely to endanger an aircraft, or any person therein...' and the penalty provided is a maximum fine of £400 or imprisonment for two years, or both. Nor is the State prevented from bringing more serious charges against a pilot if the circumstances warrant. To cause the death of others by flying in a manner which was reckless or criminally negligent could well bring about a prosecution on a charge of manslaughter, for which (in England) there is a possible punishment of life imprisonment.

This tragic state of affairs has in fact already occurred in Greece. In October 1972, a Japanese-built type YS 11A aircraft on an internal flight crashed into the sea off Athens, with the loss of 37 lives. According to the operators of the aircraft, the Captain allowed his co-pilot to let the aircraft down from a specified minimum altitude of 880 ft, until the aircraft struck the water. Because the Captain was responsible for all actions on board, he was therefore held responsible for the accident. In addition, it was suggested that the Captain abandoned his aircraft and passengers to save himself. It seems that in Greek Law there is a presumption of guilt until innocence is proved.* On the strength of the suggestions made, and a questionable report from an unreliable flight recorder, made by personnel not properly qualified to interpret the information, the Captain was arrested and charged with manslaughter. He was subsequently acquitted of the charge;† but that is not to minimise the criminal dangers attached to aviation.

All countries involved in civil aviation have their own statutory controls. The statutes vary from country to country, but they do have a great deal in common. The general principle lying behind the regulations is to achieve the safety of the general public, and in particular the travelling public, by (1) the control over the operational procedures of the aircraft, (2) the training and licensing of pilots and aircrew members to ensure a high standard of competence, and (3) the punishment of those who do not conform to the regulations. Although only the British Air Navigation Order is examined, it can be said with confidence that similar provisions will be found in the controlling statutes of other nations.

---

* Not only in Greek law: see Chapter 1, pp. 27, 57.
† In Athens, 27 September 1974.

(1) *Control over operational procedures*

The order contains 23 pages and 37 articles* detailing the duties of the operators and the crews in the performance of flying the aircraft. It is not proposed in a work of this nature to go into all the details of the order, but it suffices to state that they enforce good airmanship. Article 31 of the order is worthy of quotation as an excellent illustration of the standards demanded.

> 31. The commander of an aircraft registered in the United Kingdom shall satisfy himself before the aircraft takes off—
> (a) that the flight can safely be made, taking into account the latest information available as to the route and aerodromes to be used, the weather reports and forecasts available, and any alternate course of action which can be adopted in case the flight cannot be completed as planned;
> (b) that the equipment (including radio apparatus) required by or under this Order to be carried in the circumstances of the intended flight, is carried, and is in a fit condition for use;
> (c) that the aircraft is in every way fit for the intended flight, and that where certificates of maintenance are required by Article 9(1) of this Order to be in force, they are in force, and will not cease to be in force during the intended flight;
> (d) that the load carried by the aircraft is of such weight and is so distributed and secured, that it may safely be carried on the intended flight;
> (e) in the case of a flying machine or airship, that sufficient fuel, oil and engine coolant (if required) are carried on the intended flight, and that a safe margin has been allowed for contingencies; and, in the case of a flight for the purpose of public transport, that the instructions in the operations manual relating to fuel, oil and engine coolant have been complied with;
> (f) in the case of an airship, or balloon, that sufficient ballast is carried for the intended flight;
> (g) in the case of a flying machine that, having regard to the performance of the flying machine in the condition to be expected on the intended flight, and to any obstructions at the place of departure and intended destination and on the intended route, it is capable of safely taking off, reaching and maintaining a safe height thereafter, and making a safe landing at the place of intended destination; and
> (h) that any preflight check system established by the operator and set forth in the operations manual or elsewhere has been complied with by each member of the crew of the aircraft.

This article lists requirements with which any pilot with a sense of

---

* The articles quoted refer to the Air Navigation Order of 1974.

good airmanship would naturally comply. Failure to comply now be-comes more than poor airmanship; it becomes a crime and renders the pilot liable to prosecution.

Article 43, on the other hand, is a very wide all-embracing pro-vision. It states: 'A person shall not wilfully or negligently act in a manner likely to endanger an aircraft, or any person therein.' Equally, Article 44 states: 'A person shall not wilfully or negligently cause or permit an aircraft to endanger any person or property.' The Order does not define the word 'negligently' as used in these two articles, so it is presumed that the common law definition for the tort of negligence applies.

Probably the greatest control over safety is provided by Article 6 which requires that aircraft over 2300 kg maximum all-up weight being operated for the purpose of public transport, must operate in accord-ance with an Air Operator's Certificate. The certificate is issued by the governing aviation authority. The authority lays down requirements for the guidance of the applicants for the certificate but compliance with the requirements is no guarantee that the certificate will be issued. An Air Operator's Certificate will be issued only if the authority is satisfied that the aircraft will be flown safely and in accordance with good airmanship. There are powerful provisions in force to permit in-spection and supervision of the operation, and even though an operator does not contravene any statutory provision or regulation, if the standard of operation is considered inadequate, the authority is em-powered to withdraw the certificate and stop all flying.

The Air Navigation Order clearly recognises the dangers of pilot fatigue and, following the recommendations of the Bader Report (Chapter 1, pp. 28–33), has been amended to itemise specified duty periods which must not be exceeded. However, Article 20(8)(a) also puts the onus on the pilot: 'A person shall not be entitled to perform any of the functions to which his licence relates if he knows or has reason to believe that his physical condition renders him temporarily or permanently unfit to perform such function.'

This provision has not been judicially defined in the courts, but it is argued that a pilot will be committing a criminal offence if, for what-ever cause, he flies an aircraft knowing that, by reason of fatigue, he is unfit to do so.

Mandatory occurrence reporting. A new concept has been added to the UK legislation with the introduction of 'mandatory occurrence reporting'.

The object is to improve safety by creating a fund of statistical knowledge on experiences that may have led to an accident or incident. The intention is good and is heartily endorsed by most persons involved in aviation safety, for all too often, safety experts only hear of problem areas after damage has been done. Mandatory reporting should highlight tendencies or dangers before incidents or accidents happen. To achieve this, crew members, operators, maintenance men and manufacturers are compelled to inform the aviation authorities of any reportable occurrence. A reportable occurrence means:

(a) any incident (not being an accident...), and
(b) any defect in or malfunctioning of the aircraft or any part of the aircraft or its equipment,

being an incident, malfunctioning or defect endangering, or which if not corrected would have endangered the aircraft, its occupants, or any other person.

It requires those who have experienced a reportable occurrence to make it known to the authorities. Herein lies a serious problem area because those to whom the regulations apply must confess their errors even though no damage has been done or person injured. Such confessions might render the confessor liable to prosecution or disciplinary action, whereas silence would have allowed the mistake to escape detection. In Australia similar legislation has been in force for a number of years. Its introduction into the UK is being watched most carefully to ensure that no abuse is made of a pilot's frank self-confessions. Much depends on trust between the pilots and the authorities so that punitive action is taken only against gross errors or carelessness.

On the other hand, failure to report a reportable occurrence renders the reticent person liable on summary conviction to a fine of £400, or on conviction on indictment, to imprisonment for two years or fine or both. Time will show how this new legislation settles down into UK air law. Because of its far-reaching implications, this legislation will be kept under constant review by the authorities, in order to resolve any anomalies or difficulties which may develop. It seems that the system is working in Australia but, as yet, the USA have restricted themselves to mandatory reporting of *defects*.

(2) *Training and licensing of pilots*

To ensure a high standard, the training of pilots must conform to those requirements which lead to the grant of the licences. Little need be said regarding the possibility of criminal activities in the training process, where the only crime worthy of note would be in cutting it short. Pilot error is not an uncommon occurrence during training, however, for both the *ab initio* private pilot, and the experienced commercial pilot carrying out advanced training on new types of aircraft, are bound to make mistakes. The commercial pilot, in particular, is expected to handle the aircraft under the most adverse conditions: for example, landing without flaps; take-off with one engine inoperative; or 'two-engine out' landings.* This calls for considerable skill and the skills must be learned. But unless there has been a reckless disregard for the safety of others it is difficult to envisage a criminal prosecution following the inevitable errors. A prosecution is more likely, of course, if an operator tries to pass off his pilots as so qualified without the training.

(3) *Punishment of those who do not conform to the regulations*

The proof of guilt required for offences against the Air Navigation Order is outlined in Articles 84 and 85.

> *Article 84:* Any person who fails to comply with any direction given to him under this Order or any regulation made thereunder shall be deemed for the purpose of this Order to have contravened that provision.
>
> *Article 85:* (1) If *any* Provision of this Order or any regulation made thereunder is contravened in relation to an aircraft, the operator of that aircraft and the commander thereof, shall (without prejudice to the liability of any person under this order for that contravention) be deemed for the purpose of the following provision of this Article to have contravened that provision *unless he proves* that the contravention occurred without his consent or connivance and that he exercised all due diligence to prevent the contravention.
>
> (2) *If it is proved* that an act or omission of any person which would otherwise have been a contravention by that person of a provision of this Order, or of any regulation made thereunder, was due to a cause not avoidable by the exercise of reasonable care by that person the act or omission shall be deemed not to be a contravention by that person of that provision.

It will be seen that in Article 84, the person prosecuted needs to show good cause on his part for refusing to comply with the direction given to him. In Article 85, it will be noted that the accused are being required to prove their innocence.

One of the general defences against a criminal charge is that of necessity. This defence applies when a person is put in a dilemma situation and the choice open to him is either to commit a crime, or

* See Captain Bressey's observations on training accidents, Chapter 1, pp. 43–5.

create a greater evil. Dr Glanville Williams put the position thus: 'It is for the judge to decide whether, on the facts as they appear to the defendant, a case of necessity was in law made out, and this in turn involves deciding whether, on a social view, the value assisted was greater than the value defeated.' It is therefore not a crime for a prisoner to escape from a burning jail, or for a motorist to exceed the speed limit in an attempt to save the life of a passenger urgently in need of hospital treatment. Although there do not appear to be any reported cases where this defence has been applied to aviation cases, it is not difficult to visualise a suitable example; for instance, exceeding flight time limitations to overfly, or escape from, a country in which serious civil disturbance endangers the safety of the passengers or aircraft. There are, however, limits to the defence, as was so well illustrated in the tragic but famous case of *R. v. Dudley and Stephens* (1884); in which shipwrecked sailors killed and ate one of their party to avoid death to themselves by starvation. The court refused to accept that necessity excused the murder. Although necessity is usually referred to as a defence, the burden of proof remains on the prosecution.

The defence of necessity does *not* apply if the dilemma situation was brought about by the accused person. If the circumstances leading up to the emergency are unavoidable, then the defence *does* apply. However, if the emergency situation *could* have been avoided by the recognition of the danger, and adequate contingency made to avoid it, then the defence does not avail the accused. The most recent case illustrating this point occurred in *Higgins v. Barnard* (1972). The facts were that a motorist considered himself too tired to continue on a motorway, pulled onto the hard shoulder and stopped with the intention of sleeping before continuing his journey – contrary to the Motorway Regulations. When charged with the offence, he pleaded necessity.

Commenting on the case, the Judge, Lord Widgery, said: 'An element of suddenness is not essential ... In my judgement, in the context, an essential which has to be shown is that the driver in question was on the carriageway of the motorway, and got himself on to the carriageway in circumstances in which the danger alleged to constitute an emergency was not apparent to him at all. The next thing to show is that something supervened; not necessarily some sudden exigency at the moment, but something which rendered it unsafe for him to proceed to the next turn-off point. Given those facts it seems to me that a man should be able to plead emergency.' The facts of this particular case, however, showed that the driver was *already* in a state of fatigue before he entered the motorway; and because he was aware of the potential danger, the defence was rejected by the court.

The criminal law as it applies to pilot error is not something that the average competent pilot need fear. The aviation authorities have the teeth to the legislation, but unless there is blatant disregard for the law, these authorities are unlikely to bite. This is not to say that international pilots should not keep a very watchful eye on such of their responsibilities as may be challenged in criminal law; for where there is considered to be cause, the possibility of prosecution exists, and – as witness the case of the manslaughter arraignment quoted on p. 237, may be pressed to a degree which should not be underestimated. The hazard is confirmed in the findings of a survey carried out by the International Federation of Air Line Pilots, wherein there is evident cause for disquiet on the subject of the legal status of the aircraft commander. In Chile there is imprisonment in cases where previous circumstances give cause for presumption of the guilt of any crew member. In Colombia, in the event of an accident, the Civil Aeronautics Department withdraws the licences of crew involved in an accident – as a first measure; while New Zealand pilots seconded to internal operations in Japan reported that any incident attracted automatic police detention of the crew.

It is in this unhappy climate that international pilots must exercise their professional skills. It may be that a British or American pilot flying within his own state can relax in the comfort of knowing that his familiar environment offers him justice. That peace of mind could be misplaced in certain parts of the world.

Manufacturers and government bodies

A number of crashes have been caused by structural failure of the aircraft, or by faults in design, manufacture or maintenance. Although this chapter deals mainly with pilot error, the position of the manufacturers and regulating bodies plays a significant part. The manufacturers are under a duty not to create anything obviously dangerous. To what lengths need they go to ensure safety?

There is a duty of care imposed by both Common Law and statute. The English statute is the Civil Aviation Act 1947. Other states have similar legislation, but in general, the standard of the duty is that of reasonableness as already discussed. That such 'reasonableness' does indeed govern the design and manufacture of aircraft and their components is amply illustrated in Chapter 3, wherein John Allen has shown not only the deep consideration given to the factors involved, but in addition, the essential need for some element of compromise. But if the deliberations of risk and hazard against cost are reasonable, then the manufacturers should be safe from prosecution and civil action.

Before an aircraft may fly commercially, too, it must receive certification from the State regulating authority. These are independent bodies and the receipt of their certification goes a long way towards establishing that the manufacturers have weighed the costs and risks with the reasonableness required of them. *This is not, however, conclusive proof that they have fulfilled their duty of care.*

The regulating authority in the UK is the Air Registration Board. This body is also under a statutory duty to see that aircraft are as safe as reasonable diligence can make them, and at law, the position of the ARB is similar to that of the manufacturers, in that they must show the same standard of care.

An example of breach of care as owed by a regulating authority occurred in *Dutton v. Bognor Regis Urban District Council* (1971). In this case, the council were bound under bye-law regulations to inspect building sites to ensure that building operations conformed to the required standard. One such building did not conform to the required standard and the building inspector failed in his duty to notice the defect. The building subsided causing loss and damage to a subsequent owner, who had himself purchased from a previous owner. He could not sue the builder as there was no contractual relationship; nor could he sue the innocent vendor as he had made no warranty as to the condition of the house. The unfortunate owner therefore elected to sue the Council for their inspector's breach of statutory duty and was successful in his action.

This same standard applies to the ARB. If they fail in their duty to maintain standards of aircraft manufacture, they too might be sued as defendants. In the USA, it is not uncommon for the Federal Aviation Administration to be made defendants in air crash cases where there is reason to believe that faulty design or maintenance was responsible.*

This duty extends further, and it is necessary to examine the position when the manufacturers and regulating bodies receive warnings from the pilots and operators of difficulties or dangers that become

* See pp. 246–7 for FAA involvement in major case.

apparent after an aircraft enters service. This is not uncommon, since aircraft are in a state of development all through their working life. Certainly, warnings must not be ignored, and in practice seldom, if ever, are disregarded. But clearly, there is a difference between ignoring a warning, and not taking remedial action. What *is* vital is that the persons concerned should take careful and reasonable deliberation as to whether anything should be done.

As an example, the altimeter setting knob of a certain aircraft proved to be slow and awkward when setting QFE (so that the instrument showed zero when the aircraft landed), for high level airports. Although pilots had complained of the problems for years, the operators, the ARB and the manufacturers had arrived at the conclusion that this was a mere nuisance which did not warrant the cost of correction; all that was required of the pilots was a little more time to effect adjustment.

There the matter rested until a Comet aircraft making a night approach to Nairobi airport (5327 ft) struck the ground some 10 miles short.* Fortunately its landing gear was down and it struck level ground between an outcrop of boulders and a ravine. The pilots applied power and got the aircraft airborne again and flew on to the airport – a most remarkable escape without injuries or serious damage. The enquiry showed that the pilots were under pressure at the time and the incon- venience of the altimeter setting knob had contributed to an incorrect QFE of about 100 millibars – or nearly 3000 ft. True, the pilots were in error, but what of the warnings? Almost immediately afterwards, fast winding cranks were fitted to the setting knobs.

Was the hasty remedial action an admission of fault on the part of the authorities, or an acknowledgement that their former deliberations had not adequately assessed the degree of risk? This is a matter for reasonable men to judge. The former postulation would imply liability; the latter might not.

Other examples of this problem include the case of the cargo door of the DC 10 type aircraft. It was known that the cargo doors were difficult to lock. Was this a mere nuisance to the ground crews, or a serious danger? It could be that a real danger existed and work was in hand for modifications. But what was the cost of a hurried programme for implementing the modifications? Would it be justifiable to ground all the DC 10's until the difficulties were removed? Should the

---

* This accident is described by Captain Bressey on pp. 36–7.

modifications be compulsory; or should they be left to the discretion of the operators? Would it not be sufficient to instruct ground crews to take extra care until an economic programme was complete?

On 3 March 1974, a Turkish Airlines DC 10 airliner took off from Paris to London. Shortly after take-off the incorrectly locked cargo door blew off and 346 persons died in the resulting crash. The question yet to be answered by the enquiry and the courts is whether the questions posed had been reasonably answered.* An unreasonable decision might well be wrongful, but not a mere error of judgement. Following the crash, all the aircraft were grounded until their cargo doors were hurriedly modified. Is such rapid modification an admission of fault? Or merely a reassessment of the situation in the light of new evidence? In other words, 'the spectacles of today now see a different picture' than that of yesterday.

Sometimes a foolproof system leads to inflexibility. Following the notorious Papa India Trident crash,† everyone in aviation is well aware of the dangers of raising flaps at incorrect speeds, but to fit speed interlocks might well create other dangers of inflexibility. Pilots may still make mistakes but they are none the less highly skilled airmen and must be treated as such. Sometimes the better remedy comes from further training of pilots rather than modification of the aircraft – but this matter is considered in other chapters of this book.‡

What is pertinent to this chapter is the practical effect on the litigation that might follow. Those who have suffered loss from the DC 10 crash near Paris may well have their claims for damages limited by the exclusion clauses of the conditions of carriage. The manufacturers of the aircraft would have no such protection and if those who suffered can prove liability of the manufacturers for their loss, it is logically the more advantageous course that they should sue the manufacturers rather than the airline.§

---

* It seems this question will not receive the examination of the courts. The four defendants – Turkish Airlines (the operators), the United States Government (the inspecting body), General Dynamics and McDonnell Douglas (the manufacturers) all agreed not to contest the issue of liability; the question to go before the courts is only that of the measure of damages.
† Described on pp. 60–1.
‡ See Chapter 2, p. 79 and Chapter 3, p. 121.
§ In the event, *four* parties were regarded as liable, and were currently engaged in out-of-court negotiation fully one year after this crash.

A report in the *Sunday Times* of 9 March, 1975 said:

Turkish Airlines has offered to pay just over £4½ millions to settle out-of-court claims for damages by relatives of the 346 people killed in the DC 10 crash near Paris last March. Lawyers

Public enquiries and litigation

Hanging on the wall of one of London's most famous legal book shops is a cartoon depicting two men fighting over a cow. One is pulling the cow by the horns, while the other hauls at the tail; and in between is a lawyer wearing his wig and gown, calmly milking the beast. There is a great deal of truth in the message, but although it may make the cynic smile, it is not a complete picture.

The legal system has been with us a very long time and the more complex our society becomes, the more we need the lawyers to help us regulate it. It is worth remembering that it is the people involved in aircraft accidents who are in conflict at the hearings – not the lawyers. But since most lay people know little of the law, it is the duty of the lawyer to fill that gap and see that the laws are not manipulated in any way adverse to his client. Providing he does this, the lawyer personally 'wins' his case, even if the person he represents loses. Lawyers are men learned in the law, and it is plainly not possible for them to attain, in addition, the expertise of every specialist in all the different fields of aviation. Far better for the lawyer to concentrate on his law, and to call on such experts to supply the specialised evidence he needs to present his client's case. In the process, the lawyer may learn much about aviation, but the evidence will be that of the expert, not the lawyer.

When the Elizabethan G-ALZU crashed at Munich, it was argued on behalf of the pilot that slush was responsible (see pp. 52–3). The Canadian Authorities had evidence of slush hazards from research of 10 years earlier. Why, it is asked, was this not found and presented at the original enquiry?

Why indeed? It has been argued that this is the responsibility of the

---

representing the relatives are now talking to the DC 10 manufacturers McDonnell Douglas, one of its sub-contractors (General Dynamics Corporation)...and the US Government to decide how much each of the three should pay towards the settlement.

...It is known that McDonnell Douglas is being asked to pay 80 per cent of the damages. The final settlement may run into hundreds of millions of dollars...

An earlier report in the *Daily Telegraph* of 5 March 1975, however, claimed that McDonnell Douglas had been 'required' to pay $70 million (£30 430 000), and General Dynamics, $12 million (£5 220 000). The report added:

The United States Government, also a defendant in the suit, would be required to put up most of the balance to bring the total amount available for paying compensation up to $100 million to victims' families. The case against the Government, specifically the Federal Aviation Administration, is that it was negligent in certifying the aircraft as airworthy.

lawyer; but here, it is argued that the responsibility of the lawyer is to ask his experts to inform him of the state of knowledge of slush hazards and whether any research has been done in that particular field. It is the function of those who claim to be experts to know of such research. The lawyer, having been armed with this information, must apply his knowledge of law to ensure that such vital evidence is considered by the court of enquiry.

All hearings have conflicting evidence brought before them and those who sit in judgement must decide which of the conflicting views are the more convincing. In the case of G-ALZU, the Germans, but not the British, found the slush argument the less convincing. But if, in the future, another aircraft were to fail to take off in slush from a German airfield, who dares a prognosis of the outcome from the subsequent enquiry?

The enquiry itself is equally worthy of consideration. In the USA, it is the National Transportation Safety Board which has the dual responsibility for establishing the probable causes of an accident, and for making recommendations for the prevention of similar accidents. The Bureau of Aviation Safety may order a public hearing in the public interest and on behalf of the NTSB.

In the Federal Republic of Germany, the relevant regulations empower the Federal Minister of Transport to order a commission to be set up 'if the importance or gravity of the accident require'. And in the UK, whenever it appears to be expedient, a commission is set up to hold a public enquiry. This is not to say that every accident is not investigated by the state accident investigation branch. The commissions of enquiry are set up only when there appears to be a public need.

It is this element of 'gravity' or 'public need' that brings such unfortunate publicity upon the enquiries. It is true to say that all citizens should take note of national disasters, but the experience goes to show that the level of interest can reach almost ghoulish levels. In an enquiry however, there are two conflicting interests, namely (1) a need to demonstrate to the public that all the facts are brought forth, and that no individual or organisation, whether State or commercial, can hide their errors or responsibilities; and (2) the heartbreak and misery caused to the bereaved by the general publicity. Each State has its own approach to the problem and it is interesting to note the differences between the British and American enquiries.

In the USA it seems that the investigator in charge, and 'interested participating parties' (as distinct from any interested parties) are part of the investigating team. The Federal Aviation Act of 1958 states that 'No party to the investigation shall be represented by any person who also represents claimants or insurers. Failure to comply with this provision shall result in loss of status as a party to the investigation.'

This system makes for an investigation which is concerned solely with the facts of the accident. In contrast, in the UK interested parties are represented. This, when litigation is likely to follow, means that wide implications of the errors or causes of the accident will be pursued, as those interested will be searching for facts material to their interests. The Papa India Trident crash enquiry was a classic example, and there has been much criticism as a result. A strictly scientific approach would find greater merit in the American system, but is this really the full picture? In simple terms, the Americans search for the cause of the accident. If litigation is to follow, then that is a matter for the litigants; and they must in their suits examine further whether a person's error was an error pure and simple or a wrongdoing amounting to a tort or breach of contract. This can in essence greatly inflate the costs of the action as much of the original enquiry has to be reheard. It is possible that because of this the press and news media may be less interested, and may give less coverage to the cross examination of those involved; but there is a need for a rehearing of the facts.

The British method is more likely to put all the cards on the table, so that in the wake of the enquiry, the litigants will know better what evidence there is for or against them. Experience shows that cases are more likely to be settled out of court, as 'shadow boxing becomes pointless when the substantial facts are known. Private individuals are spared the experience of initiating litigation against usually large and wealthy organisations.' And although the publicity will have spent itself at the more dramatic public enquiry, any civil suit is necessarily a reopening of the hurts suffered from the accident. The British enquiry is said by the cynics to be for the benefit of lawyers (see Chapter 1, p. 60); but if it succeeds in reducing private litigation, then there is much to be said in its favour.

Sometimes the only matter left for decision in private litigation is the measure of damages. In this regard the American courts are more generous to the injured parties and it is often to the advantage of those claiming damages to seek a hearing in the American courts. The Americans have consistently awarded greater sums for loss of life or injuries and the precedents are there for subsequent decisions. Another reason which attracts aviation cases to the USA is that the limitations imposed by the Warsaw Convention have very restricted application, and can frequently be avoided in that country. Nevertheless, it is not every case that can be taken to the courts of America. The case to be heard must come 'within the jurisdiction' otherwise it is rejected. To come within the jurisdiction of a particular court, either the injury suffered must have occurred within the territorial limits of the court, or the person who committed the wrong must be one who lives or operates his business within those limits.

Thus, the DC 10 crash near Paris found its way into the State of California (home state of the manufacturers) for the assessment of damages after the defendants conceded liability.

Punitive damages (i.e., damages over and above the loss suffered so as to punish the defendants for gross negligence) are not recognised under Californian law, a factor that was perhaps taken into consideration in this admission of liability.

The international organisations

The very purpose of aviation is travel. Today, the main part of that travel is international and it has been shown in this chapter that the laws of one country may have little or no application to another State's legal thinking. It may be true to assert that every nation wants to see a healthy international industry, but there is little agreement on how this should be achieved. Who, for instance, should bear the loss resulting from accidents – the operator and the pilot who flew the aircraft, or the unfortunate victims? The question is probably academic to all but the insurance companies who usually carry the risks in any event; however, the international organisations do try to create some uniformity. Three such bodies worthy of mention are ICAO, IATA and IFALPA.

(1) *ICAO: The International Civil Aviation Organisation*

This is an effective body and its many conventions result in quasi-legislative edicts. For the purposes of this chapter the most relevant of these has been the Warsaw Convention. In the prevention of pilot error ICAO has probably had even greater effect in that the standards of navigational aids and facilities are laid down, and States which do not conform to these standards have pressure brought to bear upon them so as to bring them into line. The ICAO has much success to its credit, although its success is not always assured; for the edicts of the conventions are just so much verbiage unless the various nations ratify the conventions and give them the force of law by supporting legislation.

One convention signed in Rome, purported to restrict claims for damages suffered by those who were only involved in aviation by virtue of an accident. For example, the unfortunate Mr Bright, who, while eating his lunch beside the railway, was struck and killed by the falling canopy from an abandoned aircraft (p. 204). If the Rome convention had been ratified, his dependants' claim for damages might have been restricted or even excluded. This convention would have made serious inroads into national legal thinking, and the reader is probably just as well qualified as the next man to decide who, in social terms, should bear the loss. In the event, the diversity of thought proved too much for the international agreement, and the matter remains unsettled.

National sovereignty is guarded most jealously and even a convention relating to offences committed on board aircraft has met with its difficulties. The convention was signed in Tokyo in 1963 and while it would be thought that all nations would be anxious to grant powers to the Commanders to control criminal acts on board their aircraft, only 16 nations ratified. However, when the principal objects to conformity are financial, rather than political, the ICAO is more successful. Aid can always be given.

(2) *IATA: The International Air Transport Association*

This might loosely be classed as the Association of Air Line Operators, since the main function of this organisation is to regulate the uniformity of the commercial operation. IATA recognises that unfair competition and all-out undercutting of the competitors must eventually lead to a lowering of standards, and eventually, the lowering of safety standards – whatever the legislation in force. The self-discipline of the airlines is to be commended as a means to promote safety. Most big operators are members of IATA, but this is not to say that non-members do not serve a useful function in the international scene. A non-member of any association may be an irritant, but does much to prevent monopolies forming.

Apart from the strictly commercial aspects of the association, this body is interested in safety, has an influence with Government and ICAO, and can be the force behind the making of legislation.

(3) *IFALPA: The International Federation of Air Line Pilots Associations*

In the current industrial scene the unions have developed into powerful bodies. Within each nation the pilots have their own association, e.g. The American Air Line Pilots Association, and the British Air Line Pilots Association. Almost all of these associations are members of IFALPA and hold regular meetings to decide on policies, and to consider the profession as a whole. Their own legal study groups consider current legislation, and develop policies to resolve matters which may trouble them. One such matter is mentioned in the section on the contract between the pilots and the operators; namely the move to make the operators indemnify the pilots for action taken against them. A particularly active field, too, is the consideration of aircrew and passenger protection against the menace of hijacking. Nations have become very dependent on aviation, and it is within the powers of IFALPA to suspend world-wide operations, or to isolate any particular country by withdrawing the services of the pilots. IFALPA therefore has a voice and does make itself felt and heard in Government and ICAO circles.

In conclusion

Of necessity, a chapter-length review such as this must be selective. Certain of the legal implications of pilot error have therefore been examined in depth; but it has been possible only to scratch the surface of some few of the many issues which are of equal importance.

These – along with the complexities of International Law, or the virtues of prosecuting a case in one country rather than another – are matters for the practising lawyer. A major aim of this contribution will have been achieved if it has enabled the lay reader to become more aware of the significance of the point made on p. 200; namely, that it is the general law – with suitable modification by special statute – which applies to aviation.

People who take an intelligent interest in modern society naturally delve into the workings of professions outside their own. The reader who is involved in aviation, albeit as a passenger, may have learned something of the thought processes of the lawyer when considering an accident, and in his turn, may have bettered his understanding of legal responsibilities. The pilot may be disappointed to learn that the law is not as mathematically precise or as predictable as his aircraft; but equally relieved to find it neither soulless, nor entirely unforgiving towards the vagaries of man. For his part, the lawyer will have encountered some of the problems and dilemmas with which the pilot must constantly contend. It is to be hoped that future assessments of the 'reasonable pilot' will reflect this insight.

Table of cases

Table of statutes

Table of accidents

Hawker Siddeley Trident G-ARPI		18 June 1972	Staines	249
Nantes collision		5 March 1973	France	207, 236
Airspeed Elizabethan G-ALZU	CAP 318	6 February 1958	Munich	207, 247
Boeing 707 G-ARWE	CAP 324	8 April 1968	London Airport	205
Grand Canyon Collision		1956	USA	207
Vickers Vanguard G-APEE	CAP 270	27 October 1965	London Airport	213
BAC 1-11	CAP 219	22 October 1963	Salisbury Plains	207
Boeing 707 G-APFE	CAP 286	5 March 1966	Mt. Fuji	219
DC3 Dakota G-AGZB		6 May 1962	Isle of Wight	221

Appendix
Review of Typical Research Into Pilot Error

This appendix surveys present-day world activities which are directed towards a better understanding of pilot error, and to the achievement of necessary solutions.

Currently, there are essentially two approaches. Firstly, airline operators seek to identify pilot-error accident patterns and their underlying causes in order to improve training and operational procedures. Secondly, there is a more analytical attack in the laboratories of universities, aircraft manufacturers and national aeronautical research establishments. The evaluation of new equipment – or methods – is difficult for the airlines and is best done experimentally, by the laboratory teams.

Since the pilot–aircraft combination is so intermixed, work directed towards pilot error cannot always be separated from more general experiments on flight safety, and clearly the coordination of all this work is a demanding and important task. Table A1 (Agard CP 55. 6–9) summarises the scope of world research in flight safety – showing those parts which have a bearing on pilot error, e.g. man-machine relationships, human-factor studies, and improved cockpit design. The grave danger is that a great deal of expert and laudable research necessarily done in fine scale detail and based on existing cockpit arrangements can be made obsolete by some radical change elsewhere.

Operational improvements

At the 27th International Flight Safety Foundation annual seminar in November 1974, Col. Winston E. Moore of the US Strategic Air Command (SAC) reported a dramatic improvement in its safety record –

Table A1 A taxonomy of flight safety research

Problem statements	Approaches to solutions	Research objectives
Man–machine relationships	Analysis and study of accidents	**Develop devices**
Varying workload	Study a specific flight mode	Evaluate concepts
Control in turbulence	New sensor development	Display evaluation procedure
Stall, at high speed	Develop inertial instrumentation	Consider pilot's viewpoint
Stall, at low speed	Experiments on simulators	Improve information transfer
Interpreting instruments in turbulence	Simple laboratory simulators	
Fatigue, vigilance problems	Studies using full-scale simulators of air force	**End results**
Mutual understanding of responsibilities	Studies using full-scale simulators of industry	
Design-induced pilot error	Studies using centrifuge simulator	**Reports**
System goofs	Studies using simulators of airlines	Fixes of existing displays
Display failures	Experiments in aircraft	Angle of attack device
Situations pilot is not aware of	Experiments in variable stability aircraft	Improved voice warnings
Inattention	Use of total in flight simulator	New head-up display with corrective action display
False acceleration cues in the cockpit	Study turbulence effects on crew	Flying trainer/laboratory (TIFS)
Situations pilot is aware of, but needs help	Develop new display concepts	Pipe dreams
Poor instrument response	CRT displays	Gust sensor and display
Loss of warning indicators	Malfunction displays	Go–no-go computer
Structural integrity	Full use of audio display	P.I.O. predictor
Fire safety	Predictor display	Inertial instrumentation
General warning for malfunction emergencies	Corrective action display	Predictor computer
Take-off problems	Total revolution in cockpit instruments	Totally new cockpit
Go–no-go decision basis	Study computer applications	Theory of display evaluation
Control at rotation	Predictor/monitor	Improved warning system and display
Malfunctions during take-off	Go–no-go studies	Changes in training procedures
Noise abatement control	P.I.O. predictor	
Climb-out	Human factors studies	
Cruise	Study training methods	
Upset	Conduct interviews	
Unstart	Analysis of workload	
Loss of control due to speed change	Pilot reaction studies	
Rough air-upset problem	Disorientation experimentation	
Descent	Human factors in aircraft accidents	
Collision avoidance	Transfer from training	
Landing	Transfer from piston experience	
Safety in approach and landing	Transfer from military to civilian	
All-weather landing	Evaluation of displays	
Control at landing speeds	Head-up VS. head-down	
Glide path control	Command VS. situation	
Touchdown in crosswind	Inside-out VS. outside-in	
What are the landing cues?	Evaluation studies of existing and proposed displays	
Use of angle of attack during descent	Integrated displays	
Recovery lag during descent	Use of angle of attack indicator	
Terrain avoidance	Use of g meters	
International standards for safety	Use of tape displays	
C.A.T.		
Improved flight recorders		

from 200 major accidents per year in 1950 to zero in 1974. Of all the many improvements responsible for this dramatic change, he singled out the 'military integrated flight-crew concept' which achieved 'levels of proficiency and performance far beyond the expectations of social and behavioural scientists'. This practice was proved in the US space programme where 3-man Apollo moonflight teams trained and operated together for many years. If one man fell sick the whole team was changed.* This procedure is not normal in civil aviation and would probably be hard to operate and certainly more expensive.

When Japan Air Lines experienced a series of major accidents pointing to loss of overall performance on the flight deck and procedural irregularities, a Human Factors Study Group was set up to overcome the resulting crisis of confidence. Its leader reported that 'the deeper we went the more we realised that "pilot error" is no final answer and was not going to give us even an indication of how to prevent similar accidents in the future'. Crew and management met at working lunches and a confidential questionnaire probed the nature and occurrence of incidents in different phases of flight. A major conclusion was that "crew-coordination – where the error of one crew member is picked up by another – is more important than any number of warning devices'. This looks very similar to the conclusions of Col. Moore of the SAC.

The US Navy Safety Center at Pensacola reported at the International Safety meeting a correlation between pilot error and 'life changes' and illness. Life change units (LCU) were defined to account for health changes, illness and deaths of close friends or relatives, financial, domestic and emotional worries, and alterations in work responsibilities. It was found that accidents to very senior and experienced pilots could be correlated with high LCU ratings. The object of this novel approach is to retire pilots when they get a high LCU score before the pilots' impaired performance leads to a fatal accident.†

Other analysis of airline crews is seeking character patterns pointing to accident proneness. Recent work along such lines has pointed to the large part played by pride – professional, personal or false – which

* The 'military integrated flight crew' was a feature of Britain's Bomber Command life during the 1939–45 war. Like thousands of readers with this background, the Editor can testify to the important and positive effect of this system on crew morale and motivation.

† But this approach must rely on a continuing knowledge of the pilots' moods and circumstances. Knowing the consequences, he may well choose to be reticent, and/or to withhold 'key' information. There can be no 'mandatory reporting' in private life. See also pp. 83–6, p. 240.

could lead some pilots to continue a take-off or landing when cold reasoning would deduce the risk was really too great (see Chapter 2, p. 82).

Another laudable approach has been made by Boeing, the prime constructors of world airliners, in order to reduce pilot errors in airlines purchasing its new equipment. At their training school at Seattle there are 50 full-time pilots each with over 10 000 hours of airline flying. Customer pilots are first introduced to the new aircraft's cockpit in the Boeing simulator, and then passed to a special test airfield for flight indoctrination. This service comes as part of the purchase of the airliner and 4500 pilots have been trained. Furthermore, Boeing pilots accompany customer pilots during the first month of fare-earning service of a new transport aircraft, and longer if requested. By this far-ranging safety measure the designers and constructors of these airliners ensure that their equipment is correctly operated, and they can monitor at first hand any deviation in piloting skills or procedures. Clearly, pilot error is so pervasive a phenomenon, transcending many special disciplines and practices, that it is only by such initiatives as these that it can be kept in check.

Research in experimental laboratories

There are essentially four techniques at present in use:

(a) Specific tests in simulators having very realistic cockpit equipment and complex motions.
(b) Confirmatory flight tests, often in specially adapted aircraft whose characteristics can be altered by intention.
(c) Simpler 'ergonomic' tests of particular elements of the piloting task. An example is the effect of different levels of vibration on the pilot's visual performance.
(d) Mathematical formulation and study of some aspects of the man–machine relationship.

The USA, UK, Germany, Holland, France and Sweden are active in most of these techniques. The agencies conducting such tests include in the USA, for example, the National Aeronautics and Space Administration, the National Transportation Safety Board, the Federal Aviation Administration, various schools of aviation medicine at universities, the Air Forces, and a multitude of ergonomic, human factors, behavioural science and environmental departments at universities, colleges and in industry. The nature of the task is so complex that no overall national programme can be formulated, and the actual progress, the correlation between many slightly differing tests often baffles the expert, let alone the outsider.

Examples given here show two important aspects of the US programme, which is undoubtedly the most extensive of any country. Firstly, the scope of a programme at one centre, and secondly some detailed results of a pilot-error study which has developed some new diagnostic procedure.

Professor Stanley Roscoe of the Aviation Research Laboratory of the Institute of Aviation of the University of Illinois at Urbana-Champaign has built up an impressive record of accomplishment in human factors research for aerospace. Reproduced below is a summary of the integrated programme of his laboratory from which he identifies six major areas of work

(a) Behavioural engineering.
(b) Selection and training of pilots.
(c) Improvements in transferring information into the cockpit in a form useful to the pilot.
(d) Follow pilots from training through their careers.
(e) Measurements of pilot error and residual attention in navigation flights.
(f) Ascertaining the relevance of the motion in cockpit simulators.

Aviation Research Laboratory Integrated Programme Summary

(1) Conceptual models of manned system operations
(2) Complementary process of behavioural engineering and the selection and training of pilots
    Function analysis and allocation
    Display design
    Control design
    Procedure design
    Performance prediction
    Synthetic flight training
    Performance assessment
    Retention of flying skills

(3) Behavioural engineering research
    Function allocation
    The reorganisation of manual flight control dynamics and the principle of performance control
    Waypoint storage capacity in computer-assisted area navigation operations
    Display design
    Spatial orientation and the principle of display frequency separation
    Essential visual cues for ground-referenced flight control
    Bias errors and variability of distance judgements with imaging displays
    Radar target detection and the principle of visual time compression
    Control design
    Manœuvering flight performance control
    Interactive control displays
    Programmed experimental control techniques
    Procedure design
    Variations in cockpit workload and blunder rate through flight procedure simplification

(4) Pilot selection and training research
    Performance prediction
    Changing-task versus changing-pilot models
    Attention sharing and the reordering of task priorities
    Saturating workloads and the measurement of residual attention
    Operational validation
    Synthetic flight training
    Analysis of training objectives
    Incremental transfer effectiveness
    Incremental cost effectiveness
    Fidelity of simulation
        Transfer as a function of the motion cue environment
        Transfer as a function of the visual cue environment
        Transfer as a function of the procedural cue/response environment
        Computer-assisted cognitive flight training
    Performance assessment
    Reliability and predictive validity of ground-based flight checks as a function of the motion cue environment

Observer–observer and ride–ride reliability of airborne flight
checks during the course of training
Indices of desired performance and the selective measurement
of actual performance
Skill retention and refreshment
Dependence of procedural, decisional, and perceptual–motor
skills and residual attention upon recency of flight experience
Refreshment of decaying skills
Innovations in flight training
Automatically adaptive training
Computer-assisted instruction
Side-task loading
(5) Methodological considerations
Measurement of flight performance
Standardisation of experimental instructional technology
Differential dependence of display order of merit, pilot perform-
ance levels, predictive validity of flight checks, and transfer of
training upon simulated motion cue environment
Response surface experimental design modifications
Application limits of response surface methodology

An analysis of pilot-error-related aircraft accidents was reported in
mid-1974 by the Department of Aerospace and Environmental
Medicine of the Lovelace Foundation at Albuquerque.* The study
investigates over 200 major accidents to air transports which occurred
between 1958 and 1970 and for each identified over 350 data variables.
From these, 37 were selected because they were more likely to have
contributed to pilot error, and a further 11 which were chosen char-
acterised the pilot's previous record. The 48 data variables used are
shown in Table A2. Numbers were assigned to different conditions of
each variable and the data were then assembled in an aircraft analysis
data sheet.

The next step was to assemble all the numbers (for all the variables
for all the accidents) in a table printed by computer (Table A3). The
48 characteristics are numbered across the top of the table and thus
the horizontal rows give the data for separate accidents which are
numbered in the first column on the left. Thus, for example, accident
No. 17 (017) gives a '3' for characteristic No. 08. This indicates that

---

* Kowalsky, N. B. *et al.*, *An Analysis of Pilot Error Related Aircraft Accidents*. Lovelace Foundation for Medical Education and Research, Albuquerque, New Mexico. NASA-CR2 444. 1974.

the altitude of the airport was between sea level and 100 ft. At this stage there is no recognisable pattern within these data and hence no deduction can be made from what is in fact a very careful and thorough analysis of a great deal of information. Subsequently a new technique, cluster analysis, was introduced, and, using logical processes and comparisons and the dependable help of the ubiquitous computer, this method eventually regrouped the elements of Table A3 until a pattern emerged.

Table A2 Accident characteristics and variables

00 Number of engines
01 Time of occurrence
02 Type of accident (1st)
03 Phase of operation (1st)
04 Condition of light
05 Type weather conditions
06 Type instrument approach
07 Airport proximity
08 Airport elevation
09 Runway composition
10 Runway condition
11 Runway lighting
12 Runway length
13 Type of terrain
14 Pilots involved
15 Total flight time (1st)
16 Total flight time (2nd)
17 Hours in type (1st)
18 Hours in type (2nd)
19 Pilot age (1st)
20 Pilot age (2nd)
21 Pilot at controls
22 Sky condition
23 Ceiling
24 Visibility
25 Precipitation
26 Obstruction to vision
27 Relative wind component
28 Temperature

29 Wind velocity
30 Approach lighting availability
31 Pilot time last 24 hours (1st)
32 Pilot time last 30 days (1st)
33 Pilot time last 90 days (1st)
34 Duration of this flight (1st)
35 On duty time (1st)
36 Rest period prior to flight (1st)
37 Pilot time last 24 hours (2nd)
38 Pilot time last 24 hours (FE)
39 Pilot time last 30 days (2nd)
40 Pilot time last 30 days (FE)
41 Pilot time last 90 days (2nd)
42 Pilot time last 90 days (FE)
43 Duration of this flight (2nd)
44 Duration of this flight (FE)
45 On duty time (2nd)
46 On duty time (FE)
47 Rest period prior to flight (2nd)
48 Rest period prior to flight (FE)

The 'clustered' data are shown in Table A4. Note that the order of the accidents has changed and the columns of characteristics have also been rearranged, both apparently in a random way. But the clustering process has transformed the body of the data into a recognisable pattern. For example there is now a preponderance of 1s for characteristic 7 – which identifies accidents on an airport. Further groupings are brought to light by the use of colour printing of numbers and statistical methods which show correlations between causes and effects.

From this analysis the relative importance of factors becomes clear and these are summarised in Tables A5 and A6. In Table A5 are critical conditions which are shown to be more frequently present in a pilot-error accident. In Table A6 the vital importance of correct decision taking and occurrence of errors in this process are brought together.

Now the strength of this study becomes clearer – from a large sample of accidents a mass of *likely* factors has been assembled, initially in overwhelming complexity, but later yielding patterns which could then be interpreted to show important issues and priority causes. Never-

Table A3 Data on 56 non-training/non-midair accidents

Characteristics

```
          0000000000111111111122222222223333333333444444444
          012345678901234567890123456789012345678901234567
          -------------------------------------------------
     001  121122313211383231121332111192232322333232329221
     002  132223213222383322232123323411112233232232222332233
     003  122223213222383322122322323211133322323339222233
     005  132223223122213331132323323492121933322999933333
     006  131112213112383321132312131129219212223299221122
     007  121123913189383332122123323481139933333999933333
     008  131112313131283231122311211299229931122999331111
     009  131112313211283333131313213499122222322323 9222233
     010  122223212299283323222231332339993233332399113331323
     011  111112313111333222123311221121219911311199 99111133
     012  132123213129283232222323323411929231223993 9331122
     013  121123213122293322132113323422133233333229 3333333
     014  232213221199313331122313323421239933323 9999332323
     015  131112313221383323232112313432231122333122 2222233
     016  122323933199223331122922323299931232339 2929392939
     017  131112313211283322332333221243223232323 22233222232
     018  131112313221383331121111211222233922333 9999222233
     019  122223213229283333231123323412911921211 2299221122
     020  232212213111283211121311211292229921322 9999221133
     022  131113212121383311121123323322299339339 33933333333
     023  121123313229283219921123323481939911333 99999111133
     026  132212322111323212121332121792113222223 92999222929
     030  311112313111213212199132113432239233339 9912393939
     031  222222313119313212221312111492229322329 9919292939
     032  222222323191243211112392121121223332299 29999999999
     033  122123313932313222313112323411933332333 233332933
     035  131113313111223232221133313111211231319 1929391939
     037  331123213121132312211112313491229331329 9939391939
     040  331123913232313232119123313491929122329 99939292939
     041  322211323231113332122133121121929931229 9999391929
     042  122222329911933222233912319199221131229 9999391929
     043  332113229931923112211323323491139933339 9999393939
     046  221123311199383211111313313422939322331 9999222132
     049  131422438888888333122231331382882333232 21113332233
     050  111413413119383333232112393922829221222 9923221122
     051  131413412221283313221113323231893323339 22332233
     053  111512412211383222222312211931839932333 9999332233
     054  212412412221383212122312313232822331122 2232311111
     056  122412422131313333322133221331833933333 3399333333
     057  232312438888883323232132221182183933333 3992323333333
     060  231513412111383212211323323911819932111 9999332211
     064  211413413211183232222112321911839923333 9999332233
     067  132423413222383323232113323221811131211 11111331122
     069  131122913911383322232312219999931229399 9999229933
     070  113312888888888322112311999989983992233 33999922 2233
     071  121122313921233322232111129999929922322 9999222233
     078  332212838888882232112233111388883933333 99939393939
     080  111413813111383312122293313221839233339 9921333333
     083  334423413199313212221223323992839232333 99929332233
     084  332212129999992333322131231129123223233 91929392939
     085  132412412111313223323211311431821122322 1219222233
     055  322412412221213331121312211432822233129 3929393919
     074  121123211922383331199313323111999233999 9933333399
     079  393313438888882333222231232389882322232 91919292939
     086  231123213122113221222323323412129939399 99999999999
     092  133323438888888333333321391138888299329 229999332299
```

Accidents

Table A4   Clustered data on 56 non-training/non-midair accidents
Lines demarcate major clusters and subclusters.

```
                              Characteristics
              0020021120022112431142332022431333314340344440101
              7024251899310076556368204173199173827688434025564
              --------------------------------------------------
     018      11111111|2211121122382232232199233933333222992333
     008      11111111|9113231111381992232199222922111333392233
     009      11111111|9213113322382921234333322222333322292333
     035      1111111|211111123113291123313212112923393339993233
     053      1111111|2125321212228239821929923393333323333992243
     015      1111111|2221122332228231233422133313333332222222333
     006      111111|212113211311281119131293322923222322292323
     011      11111|211111331211123129221119921191333331111992233
     051      11111|21212411233221823983323292333233332333923343
     050      11111|9922141213311381293319229322223222322933343
     085      1111|2113114231312221231833411112221233382222292243
     069      111|2111299132129992899192392993392933333222992393
     071      111|21212991122292223299912919932292233333222992333
     074      111|212211913921333383199321339999293999133933323
     013      111|212212111222333293231324392333232333333233323
     067      111|22223124132331128121833231121111322233333113343
     033      111|2222319113323222191393242331333333333333233233
     010      111|2229292232933132339293233192323322332333933323
     022      11|21121121111213391832923333339293333323333933323
     007      11|221291111128233338389132439923393333333333993393
     023      11|221299121112931118189932439923393233331119932 33
     003      11|2222211223223223222821313223323333333333222993323
     002      11|2222212212223222821213343233223233333333233223323
     012      11|222292111322331128119933433922223222223333993323
     019      11|22229222211233113811193243923119122223222293323
     017      11|31111322132122222283222234232333322323222232333
     001      11|321112213111122382922121222233333231322933223 3
     080      11|211111142212333183298312329923332333333333913383
```
Accidents
```
     064      1|21112121241212122382198319299233931333333399 3243
     020      1|211211121231111111819922321992229223333222992223
     054      1|211211122432223111813283122322223231112333222243
     046      1|212119121131913221812993243991333133321222992 33
     031      1|21221922123112122119992124219222939333922999 2233
     060      1|22112121153112322182198339399111913111233399 3243
     086      1|12212222113221399219191913343992299919393399 9223
     055      1|31121112243121133319328224223222292119233999 2343
     037      1|31211221111111311313199923342392223913399399 3223
     040      1|32211211211932322319999334339122193339322999 3293
     083      1|322429221421932211299833932923323333333339993243
     030      1|3311111211191233311939211421993329233933399222 33
     042      2|1122112992931291123991232129932219922993399922 33
     005      2|12222221213231333313911334399322922333333339993
     026      2|1331211111231121221299212322922332932292222992 33
     056      2|13122131141233133313338223393923393333323333392 343
     014      2|21122991112329132331329293343992329332331333993 323
     032      2|29222211112329119214923212129212933929933999923 23
     084      2|31121921923193122399223322212332993939933999923 13
     043      2|32122121913132333129991334399133999339933999932 123
     041      2|33122111221232111319299121399222991229333999132 33
     049      3|11211828843281322382228333831222328333383331323 43
     016      3|122229191392913223299193222222333292333933399393
     092      3|1323183883128332238289813899932928999833339933 43
     057      3|22121831833182133333298238229333383338333399 3243
     079      3|3311328288332823223299383982111222298339822999 3343
     078      3|3331218188232813332298898138139233983398339992 382
     070      8|1113981983138192228299891899923393833338222992 283
```

theless, this is not a total explanation; rather, it is a link in a long chain of related research. Many thousands of copies of the report have been issued world wide so that the new evidence can influence other studies, and training programmes for flight crews–and also help designers to make safer flight decks, less conducive to these faults. Many of the points made by authors in this book show up in Tables A5 and A6; for example the devastating effect of hurrying, saturation of multiple tasks at critical moments, and inadequate training and experience.

Two significant statements were made at the 27th International Flight Safety Foundation seminar already quoted on p. 255. The first of these statements focused on the worsening trend in jet airline accidents (excluding hijackings and warlike acts). A possible cause for this was quoted as cutbacks largely affecting crew training–emphasising the careful and continuing attention which must be paid to all aspects of safety, especially as airlines adjust to higher fuel prices, the effects of escalation, changes in traffic, and the scheduling of routes flown by different types of aircraft.

Table A5 Critical condition categories

(1) Experience
  (a) Low pilot time in type
  (b) Low co-pilot time in type
  (c) Low pilot time in position (as Captain)
  (d) Low co-pilot time (total)
  (e) Other (includes: recent experience, training, flight engineer, age differences, student pilot new, crew new, student pilot dull, new airport)
(2) Distraction
  (a) Communications or traffic (excessive communications with ATC or looking out for traffic)
  (b) Confusion (last minute approach changes or other confusion)
  (c) Hurry (close departure on same runway or other hurrying)
  (d) Holding or delay
  (e) Other (includes: wake turbulence, numerous distractions, foreign student, first officer monitoring instruments, interrupted checklist, fuel burn, paperwork, poor destination weather, instructor pilot checklist, take-off position holding, ashtray fire)

(3) Crew coordination
  (a) Disagreement (disagreement on approach or configuration, or other pilot calls 'off profile')
  (b) Jumpseat occupant or other additional crew
  (c) No required altitude callouts
  (d) Pilot acting as instructor
  (e) Other (includes: loose student/instructor relationship or other interactions such as flaps without student knowing, altitude confusion, distrust first officer, thought continuing take-off, gear up without visual verification, both pilots on controls, non-compliance, confusion on who was flying)
(4) Neglect
  (a) No cross-check on ILS
  (b) Improper use of checklist
  (c) Improper rest/procedure
  (d) Other (includes: company did not revise checklist, other aircraft collision light off, ATC, Mach trim switch, engine reversing indicator lights, VOR out, clearance deviation)
(5) Air traffic control
  (a) Delayed landing clearance
  (b) Confusing radar vector
  (c) Advised of traffic
  (d) Poor, weak or malfunctioning radar or radar return
  (e) Other (includes: no acknowledgement, no advisories, vector confirmation, advisory holds)
(6) Decisions
  (a) Off acceptable profile
  (b) Institutional decisions: OK to operate
  (c) Co-pilot flying, taken over by pilot
(7) Work/rest (fatigue)
  (a) On duty over 8 hours
  (b) Minimum rest
  (c) Early morning departure
(8) Machine
  (a) Gross weight (overweight or heavy gross weight)
  (b) Simulated engine shut-off (engine failure simulation)
  (c) System failure
  (d) Other (includes: simulated rudder loss, flight director oscillation, spoiler deployment and retraction, battery

switch, 3 and 4 engine reverse, slow spool, air noise, parking brake versus mechanical failure, seat failure)

(9) Airport
  (a) Stopping problem (runway slippery, wet, slush, braking action poor, or tyre residue)
  (b) Touchdown problem (runway short or displaced threshold)
  (c) Vertical guidance problem (no approach light, approach lights out, or localiser only)
  (d) Runway hazards (upslope threshold, exposed lip or drop-off)
  (e) Other (includes: runway markings obliterated, uncontrolled airport, irregular lights, loose pavement, hilly terrain)

(10) Weather
  (a) Visibility problem (heavy rain at threshold, below circling minimums, fog, snow or haze or other visibility restrictions)
  (b) Thunderstorm influencing airport or en-route weather
  (c) Wind gusty
  (d) Other (includes: same route, weather above circling minima, en-route weather, freezing drizzle, venturi wind)

Table A6   Critical decision categories

(1) Decisions resulting from out-of-tolerance (off profile) conditions
  (a) Take-over of controls
  (b) Verbal instructions between pilots
  (c) Excessive deviation called out
  (d) Inadequate braking observed
  (e) Assistance in flight-control operation
  (f) Attempt to regain directional control
  (g) Go-around initiated
  (h) Other*

(2) Decisions based on erroneous sensory inputs
  (a) Approach continued visually
  (b) Decided profile within limits
  (c) Misleading cockpit display
  (d) Misleading navigation information
  (e) Runway/braking misinformation
  (f) Final approach/flare profile misinformation

*Critical decisions listed as 'other' in each category were miscellaneous and too few in number to list herein.

    (g) Standard operating procedure distraction
    (h) Other
 (3) Decisions delayed
    (a) Take-over of flight controls or assistance
    (b) Go-around decision
    (c) Take-off abort
    (d) Thrust lever movement
    (e) Other
 (4) Decision process biased by necessity to make destination or press-on (meet schedule)
    (a) Continue flight with equipment failure
    (b) Alter cockpit procedures
    (c) Continue with weather conditions deteriorating
    (d) Runway misinformation
    (e) Decision involved approach procedure
    (f) Other
 (5) Incorrect weighting of sensory inputs or responses to a contingency
    (a) Deviation from checklist/altitude callouts
    (b) Icing of aircraft
    (c) Disregard of cockpit displays
    (d) Traffic information disregarded
    (e) Disregard information on landing environment or conditions
    (f) Safety degradation due to training
    (g) Other
 (6) Incorrect choice of two alternatives based on available information
    (a) Left cockpit
    (b) Landed runway with unfavourable conditions
    (c) Flew visual approach
    (d) Other
 (7) Correct decision
    (a) Checked approach light level
    (b) Confirmed minima
    (c) Took over and flew approach
    (d) Other
 (8) Overloaded or rushed situation for making decision
    (a) Primary attention diverted
    (b) Aircraft power difficulty
    (c) Observed traffic and rolled aircraft
    (d) Other
 (9) Desperation or self-preservation decision

(a) Directional control or stopping problem
(b) Airborne loss of control
(c) Avoid ground contact
(d) Avoid other aircraft
(e) Other

The second statement, by Mr C. O. Miller, former Director of Aviation of the US National Transportation Safety Board, was contained in a paper on law and air safety, and stressed the need to move away from the legal concept of the 'prime cause' – recognising that this classic attitude conflicts with the fundamentally complex and additive nature of errors of many types and causes, which create the environment of an aircraft accident.

* * *

Painfully, and at such bitter cost to so many of those who had no thought but to serve the industry with pride and integrity, aviation has advanced so far; hopefully, a very long way from the sad verdicts of Ciampino and Munich and Nantes – and hopefully, too, along a path of awareness from which none can now turn aside.

The improvement in the safety of aviation, the bettered trust between all in its service, and the new methods of analysis, design and performance which will undoubtedly emerge in the future – all of those things will owe their origins to those who, in the face of so many problems and disasters, seek, not for scapegoats, but for answers. It is to be hoped that the great professional effort described here will indeed result in a more mature understanding of, and a more rational approach to, that collective failure which has hitherto been called 'pilot error'.

Index